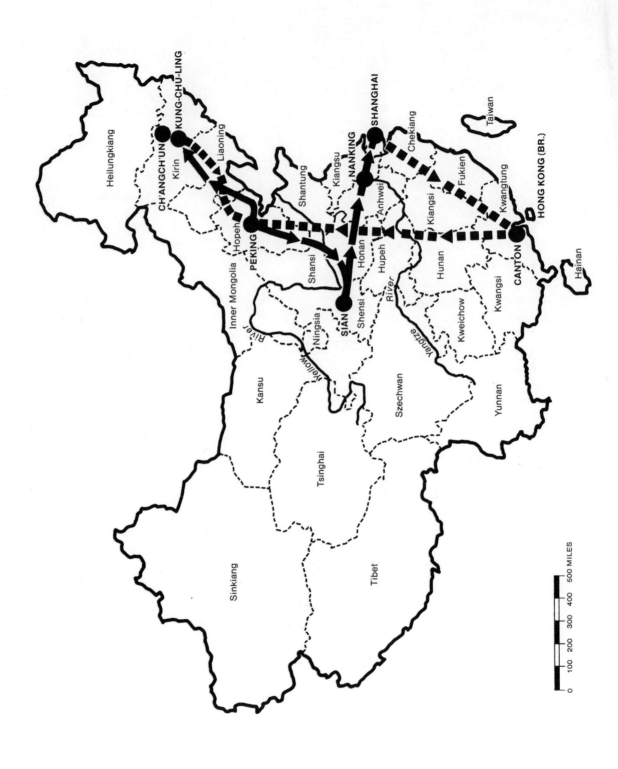

PLANT STUDIES in the PEOPLE'S REPUBLIC OF CHINA:

A Trip Report of the American Plant Studies Delegation

Submitted to the
Committee on Scholarly Communication
with the People's Republic of China

NATIONAL ACADEMY OF SCIENCES
Washington, D.C.   1975

Library of Congress Catalog Card No. 75-13564
International Standard Book No. 0-309-02348-3

Available from
Printing and Publishing Office
National Academy of Sciences
2101 Constitution Avenue, N.W.
Washington, D.C. 20418

Printed in the United States of America

CONTENTS

INTRODUCTION

The U.S. Plant Studies Delegation, organized under the auspices of the Committee on Scholarly Communication with the People's Republic of China of the National Academy of Sciences, the Social Sciences Research Council, and the American Council of Learned Societies, spent four weeks in China, from August 27 to September 23, 1974. The delegation consisted of ten plant scientists, a staff member of the Academy who served as secretary of the delegation, and a professor of Chinese history (Appendix I).

The delegation had three principal objectives. First, it was to establish contacts with Chinese authorities and scientific personnel for the purpose of initiating exchanges of information and biological materials which would be mutually beneficial to the two nations and which, hopefully, could continue and expand in the years ahead. Second, the delegation was to obtain an understanding of the status and organization of Chinese scientific and technical work in the plant science field. Third, it was to report its findings to the U.S. scientific community and to the National Academy of Sciences Committee on Scholarly Communication with the People's Republic of China.

In the organization of the U.S. team, an effort was made to include expertise on the food crops of importance both in China and the United States as well as authorities in botany, plant genetics and plant breeding, taxonomy and evolution, plant pathology, horticulture, agronomy and soils. It was considered desirable on this initial trip to send scientists with a practiced eye and with records of accomplishment that were likely to be known by Chinese scientists. Attention to the plant sciences as related to food production was considered to be the appropriate focus of the team's efforts, given the seriousness of the world food-population problem.

Most members of the delegation met in Washington on May 6-7, 1974 for orientation and to prepare a proposed itinerary for submission to Chinese authorities for their consideration. The members had found almost no up-to-date information available on the organization of Chinese science and agriculture.[1]

---

[1]Later, the delegation received copies of the comprehensive and remarkably accurate report "Making Green Revolution" by Benedict Stavis, Cornell University.

But, on the basis of reports by scientists and others who recently had been to the People's Republic of China, and on such published information as could be obtained, the delegation submitted a proposed itinerary for the four-week period August 27-September 23 (Appendix II). This was translated into Mandarin and transmitted through the PRC Liaison Office in Washington to the Scientific and Technical Association, Peking. The Chinese authorities accepted the suggested dates, and later kindly arranged for visits essentially as requested by the U.S. delegation. The actual itinerary followed is in Appendix III.

Our hosts, the Association of Agriculture of the People's Republic of China, did much to make the delegation's visit a success. First, it assigned several competent and congenial people to travel with the team: they were Professor Lou Ch'eng-hou, a plant physiologist who had studied at the University of Minnesota, and who is knowledgeable about scientific agriculture; Mr. Chou Chenning and Mr. Huang Yung-ning, who made the many institutional and other arrangements in various parts of the country; and Ms. Yang-hung, Mr. Ch'en T'ien-lin, and Ms. Hsu Chin-hua, who handled translations.

Second, arrangements were made for the delegation to see experiments in the field at provincial academies, at the production brigade level, and in the laboratories and fields of the Academy of Sciences (Academia Sinica) and the Academy of Sciences of Agriculture and Forestry in Peking. At each location, the scientists in charge described their work and answered questions by American team members.

Third, the delegation was afforded opportunities to discuss problems of mutual interest with Chinese specialists - and in small groups and at length - at several locations where major experimental work is underway.

Fourth, visits were arranged to communes, production brigades, and production teams, with leaders of the Association of Agriculture and of Academies of Science and of Agricultural Science, and with political leaders and peasants. All these opportunities added to the team's understanding of the policies, structure, and human resources involved in China's great progress in agricultural development.

Fifth, germ plasm of a number of crops was presented to the delegation to take back to the United States and to international research institutes, usually in response to stated interests of individual team members. The items received are recorded in Appendix V, as are the materials sent to China by American scientists.

Before entering China, members of the delegation kindly accepted responsibility for specific aspects of studies in China and for preparation of the resulting sections of this report, for which I as chairman wish to express my gratitude. While all team members participated in drafting the report, specific credits are in order, as follows: Entire group - conclusions; Dr. Bernard - soybeans; Dr. Borlaug - general review of agriculture in the People's Republic of China, forestry (with Dr. Creech), wheat, barley, triticale; Dr. Brady - Chinese agricultural research and extension system, rice; Dr. Burton - forage and pasture grasses, millets, turfgrasses; Dr. Creech - fruits, ornamentals and medicinal plants, botany and botanic gardens, Appendix V; Dr. Harlan - sorghum, cotton, potentials for germ plasm exchange (with Dr. Creech); Dr. Kelman - higher education in the agricultural sciences, plant pathology (plus assistance to other team members on diseases of crops); Dr. Munger - vegetables, horticulture; Dr. Sprague - corn, plant breeding and genetics, plant physiology; Mr. DeAngelis - list of persons met, as well as translations into English of lists of germplasm samples and publications received; and Dr. Kuhn - social and political factors affecting agriculture.

Finally, I should like to express appreciation to Mr. Henry Romney and his staff for preparation of 1000 copies of the handout listing members of the delegation, which were found to be exceptionally useful, and for assistance in preparing copies of this report.

Sterling Wortman

COMPENDIUM

During their four weeks of travel and study in the People's Republic of China, the U.S. Plant Studies Delegation reached a number of general conclusions.

1. The U.S. group was received warmly, and our Chinese hosts made every reasonable effort to arrange for the delegation to hold talks with Chinese counterparts on a small-group or individual basis, to see crops in the fields of institutes and communes, and for individual U.S. scientists to see places of particular professional interest to them.

Seed samples were exchanged and interest was expressed by both sides in the continuation of such exchanges. Chinese scientists at several institutes presented requests for particular germ plasm which they wish to test, and the U.S. group is to continue to present requests to China for items needed by American researchers. The climate for such exchange seems good.

2. China during recent years has focused the work of all agricultural administrators, scientists, educators, technicians, and farmers on increasing yields of basic food crops and stabilizing them at high levels wherever possible. Stabilization has been sought through improved varieties and cultural practices, massive and widespread land leveling, irrigation and drainage, and flood control efforts backed up by rural electrification, road and bridge building, provision of chemical fertilizers to supplement traditional organic types, and arrangements for marketing products at stable prices. The nation seemingly has been remarkably successful, for crops generally looked good wherever the team traveled. It must be remembered, however, that a good rainy season was underway, and that the team did not visit the driest or more remote regions. Nevertheless, the delegation was favorably impressed by the quality of farming and the appearance of most crops in many of the regions visited.

3. The system of organization of farmers into communes, production brigades, and production teams should permit rapid exploitation of further advances in science or agricultural technology.

4. No doubt, China can raise output further by applying additional existing crop and animal technology, using greater amounts of chemical fertilizer, and further improving her land and water resources. Future increases in productivity will likely be more difficult to achieve than were those of the past decade, since so much of existing knowledge which could be quickly employed is already being exploited. There appears to be greater potential for progress with corn and sorghum, the millets, vegetables, and fruits. While in all regions visited rice looked good by world standards, there is potential for further progress, particularly in breeding for disease and insect resistance. China has not taken advantage of advances in forage crop production. Wheat was not seen.

5. The Chinese policy of self-sufficiency has brought about remarkable achievements and is a source of great pride. They have proved to themselves that they can overcome great obstacles without appreciable outside assistance. The team was impressed, however, with the extent to which science interchange with other countries could be mutually beneficial. China can gain much by broadening its scientific contacts with the outside world.

6. Education and the sciences in China are being reorganized. Goals and structure of research are still in a transitional stage, and the outcome of this transition is not clear to the group. Chinese authorities point out that it was necessary to disrupt and adjust science and education - even to the point of moving whole institutions from urban to rural areas - to make them relevant to the needs of the people. At any rate, China's agricultural progress during the balance of this century will depend in large part on the ability to reconstruct scientific and educational institutions in such a way that new scientific and technological information and materials are produced in a highly effective way and, new generations of highly capable researchers and teachers are trained in substantial numbers to replace the relatively few highly trained scientists and scholars now available, most of whom are quite elderly.

7. Responsibility for most agricultural research and education has been decentralized. Many research institutions and colleges that were formerly regional or provincial components of a nationally administered system are now administered by provincial or municipal governments. While this helps to assure that the institutions will work to a greater degree in direct support of the people's communes, there is danger that coordinated research on problems of regional and national importance may not evolve. On the other hand, if China is successful in restructuring education and science so that they are highly effective yet do not lose the close ties to production efforts at the commune

and brigade level, agricultural advances in the future could be quite substantial.

8. China appears to be making an all-out effort to reduce the population growth rate. This must be done quickly, for a population increase of no more than two percent per year mandates an increase in food output equivalent to 5 million tons of grain per annum just to maintain present standards of living. Failure to reduce population growth rates now would probably create very serious food problems for the country within a few years.

The members of the delegation are unanimous in their assessment that the visit to China was highly successful.

THE ORGANIZATION OF SCIENCE AND AGRICULTURE IN CHINA

Most agricultural activities of the People's Republic of China are the
responsibility of the Ministry of Agriculture and Forestry, formed in 1970 by the
merging of several ministries including agriculture, forestry, state farms and
land reclamation, and aquatic products.

Agricultural education and research are the responsibility of the
Ministry's Bureau of Science and Technology. Agricultural research is carried
out primarily by academies of science of agriculture and forestry, and by colleges of agriculture, in cooperation with personnel of communes.

Association of Agriculture: Professional Societies

The Association is the central and most powerful "unofficial" agricultural society. Its membership is drawn from the ranks of scientists, administrators, and farmers (following the 3-in-1 principle of combining scientists, technicians or administrators, and workers or farmers).

In addition to the central Association of Agriculture of the People's
Republic of China which has offices in Peking and was our official host, there
are provincial or municipal Associations of Agriculture. Such organizations made
specific arrangements for our visits to their areas.

The various professional societies (e.g., Society of Agronomy, Society
of Plant Pathology) reportedly have been disbanded and publications suspended,
pending reorganization. They apparently will no longer have only scientists as
members, but will take in others, following the 3-in-1 principle.

The team was informed that the Association of Agriculture may soon
start publication of a general journal of agricultural research, delaying until
later any decision on disciplinary publications.

The Academy of Sciences

The Academy of Sciences, or Academia Sinica as it is called in China, is the highest, or most prestigious, scientific institution in the country. Its president is of ministerial rank - on a par, for example, with the Minister of Agriculture. Therefore in the government hierarchy, Academia Sinica is considerably above the Academy of Sciences of Agriculture and Forestry, which is under the Bureau of Science and Technology of the Ministry of Agriculture.

Like other government organizations, Academia Sinica is undergoing reorganization or "adjustment."

The group visited the Academy on August 31, and was received cordially by vice-president Wu You-hsun and a number of senior scientists of the Institute of Botany. The delegation then broke into two sections for visits to specific laboratories where scientists explained their work, present and past. Accounts of those presentations follow. Publication of the results of the Institute of Botany has continued without interruption (and the group was presented recent volumes of Acta Botanica Sinica and Acta Phytotaxonomica Sinica). Other Academy publications have been suspended, at least temporarily.

Later the delegation visited the Institute of Genetics in Peking. Reports on those visits are given in later sections on specific crops or specific disciplines.

Soil Science Institute

The Soil Science Institute of Academia Sinica is located in Nanking. We had only one hour for a very brief look at the institute which has the following divisions: soil geography, soil chemistry, soil physical chemistry, soil microbiology, soil physics, soil biochemistry, and saline soil studies.

Professor Dr. Yi Hsiung, a knowledgeable soil physical chemist, who received postgraduate training at the Universities of Missouri and Wisconsin, heads the institute. His title is associate director, the same as that held by Professor Dr. Li Ch'ing-K'uei, a soil chemist trained at the University of Illinois. Laboratories were clean and appeared to be in use. Young scientists were able to give sensible answers to questions about their work. Some good analytical equipment manufactured in China was available, including an emission spectograph, a mass spectrometer, and an electron microscope. Most of the laboratory work appeared to utilize relatively standard wet laboratory techniques common for plant and soil analyses. Except for the equipment listed above, the laboratories would not be particularly sophisticated. At the same time, the scientists seemed to know what they were doing and why.

A visit to the library showed the most up-to-date set of foreign journals that we saw on this trip. <u>Soil Science</u>, Soil Science Society of America <u>Proceedings</u> and <u>Soils and Fertilizers</u> all appeared to be up to date or nearly so. <u>Advances in Agronomy</u> seemed to be only one volume behind publication.

Scientists at this institute appeared to be as knowledgeable of advances outside China as any we visited. This may be due to its location in Nanking where there is a tradition of scholarly and scientific excellence.

Academy of Sciences

The group inquired about the location of those institutes of Academia Sinica which are related to the plant sciences. This was accomplished by going through the institutions, one by one, which are listed in the <u>Directory of Selected Scientific Institutions in Mainland China</u>, Hoover Institution Press, 1970. The following are current locations as indicated to the delegation:

Chinese Academy of Sciences	Peking
Institute of Botany	Peking
Central South Institute of Entomology	Peking
Institute of Genetics	Peking
Institute of Entomology	Peking
Institute of Microbiology	Peking
Institute of Applied Mycology[1]	Peking
East China Institute of Entomology	Shanghai
Institute of Plant Physiology	Shanghai
Institute of Biochemistry	Shanghai
Institute of (Pedology) Soils and Fertilizers	Nanking
N.W. Institute of Plateau Biology	Tsinghai

Agricultural Research and Extension System

The Chinese Academy of Agricultural and Forestry Sciences (CAAFS) is the research and extension arm of the Ministry of Agriculture. Apparently the Chinese Academy of Sciences (CAS) provides some disciplinary advisory and collaborative inputs (see Organizational Chart).

Prior to the Cultural Revolution agricultural research was administered centrally by the Chinese Academy of Agricultural Sciences (CAAS). Research was implemented through national research institutes and provincial branches

---

[1] The Institute of Applied Mycology, formerly a separate unit, has been combined with the Institute of Microbiology.

of CAAS. Some of the institutes were located in Peking, others in the provinces. Even the latter, however, were under the central administrative control of CAAS.

Those institutes of the CAAS located in Peking included the Institutes of Plant Protection, Animal Husbandry, Agricultural Mechanization, Crop Breeding and Cultivation, and Soil and Fertilizers. Located in the provinces were the Institutes of Tea Research, Veterinary Sciences, Citrus Research, Pomology, Hemp Research, Cotton Research and Sericulture. In addition there were a number of provincial branches of CAAS administered centrally along with the national institutes.

During the Cultural Revolution three significant changes were made in the organization and administration of agricultural research. First, the Chinese Academy of Agricultural Sciences was combined with a similar Academy of Forestry Research into a single Chinese Academy of Agricultural and Forestry Sciences (CAAFS). Second, the administrative control of agricultural research was largely decentralized. Third, most of the research was focused on the immediate solution of practical crop and animal production problems at the so-called "grass roots" level.

Through decentralization, most of the agricultural research institutes located in Peking were moved to the provinces. These institutes along with those already in the provinces and the provincial branches of CAAS were all placed under the administrative control of the province or municipality in which they were located. Under this greatly decentralized system, CAAFS apparently plays only a general coordinating role and is concerned primarily with national and regional (inter-province) problems and not with those at the provincial level. However, the system seems to have great flexibility and the involvement of CAAFS in a given problem area may be more direct than this general description would suggest. For example, wheat rust surveys which require inter-province cooperation are likely coordinated by CAAFS.

The Provincial Research-Extension System

Each description given of the provincial level research-extension-production program focused on the "masses," the "grass roots," and the "3-in-1" approach. Apparently everyone actively engaged in agricultural research must devote some time to work at the brigade or production team level. Yet there appears to be great flexibility in the system with no rigid uniformity from one province to another or from one commune or brigade to another.

- 4 -

There are two primary institutional units for agricultural research within a province: provincial level academies, institutes or colleges of agriculture; and small agricultural experiment stations that are
generally administered at the production brigade level though in some cases
are administered at the commune or even county level (see Organizational
Chart). The provincial level units perform applied research on the range
of crops and animal species most common in the province. To assure focus
of these units on practical farm problems, staff of the provincial level
institutes, academies, and colleges are stationed on a rotational basis at
the "basic points" at selected brigades. At any one time about one-third
of the staff would be so stationed. They also assist in the training of
technicians, farm managers and farmers.

The typical local agricultural experiment station focuses on the
major problems of the brigade and commune in which it is located. Personnel
from the following groups provide a cooperative effort to solve the local
problems: scientists from institutes, academies, and colleges from both
provincial and national levels who are stationed at the brigade; peasant
technicians with special training in applied research and extension activities; and crop production managers at the brigade level. This provides
the "three in one" approach which assures not only the scientific and
technical quality of the work but also its direct applicability to problems
at the commune (farm) level.

Research findings and plant materials developed at provincial and/
or national institutes, academies, and colleges are tested at these field
experiment stations. Also, new varieties and technologies are developed at
the experiment stations to fit local conditions. For example, inbred lines
of corn and sorghum are developed and used in production of hybrids. These
stations in turn cooperate with nearby production teams in field testing of
new varieties and technologies and in the larger scale production of hybrids
for farmer use. Experiment station personnel train technicians at both the
brigade and team levels and participate in extension activities with farmers.

Personnel at the county and commune levels normally are concerned
more with production and planning than with research. They apparently use
research findings in making decisions on production program planning,

including seed and chemical needs. They also participate in agricultural research planning, helping to determine problems or areas to be investigated.

Production teams or brigades may be involved with applied research, but are more often concerned with testing and demonstrating results obtained at the experiment stations, institutes, academies, and colleges. They produce hybrid and other seed and utilize technicians as well as scientists to train farmers to use the new seeds and improved practices.

By using the scientific talent from the provincial and national academies, institutes, and colleges to help plan and perform research and extension activities at the production brigade and production team levels, the time-lag between discovery and practical applications is shortened. Evidence available to the delegation suggests that this objective is generally being achieved. Similarly, using these same scientists to help train peasant technicians and even farmers in such science-related activities as hybrid corn and sorghum development, helps assure rapid dissemination of knowledge.

Unknown is the effect that this system (which focuses on the immediate use of existing knowledge) will have on the gaining of new knowledge. It certainly orients all scientists to practical farm problems. At the same time, it must reduce basic research efforts and makes difficult any experimentation which requires continuity of time and effort.

It has been suggested that the current situation is merely the first stage in the development of a more balanced research program which will give adequate attention to fundamental research as well as the applied research and extension efforts. In the meantime, the practical needs at the farm level will be firmly implanted in the minds of scientists at all research levels.

Kirin Academy of Agricultural Sciences

Located at Kung-chu-ling, Kirin province, the KAAS was founded in 1949 under the name of the Northeast Agricultural Research Institute. In 1959 it was renamed the Kirin Academy of Agricultural Sciences. Its purpose is to help increase crop and animal production in the province.

There are 757 employees working for KAAS of whom 255 are scientists or have science training. On a three-year rotating basis, one-third of these scientists work full time at KAAS headquarters, one-third are located at headquarters but are on call to provide technical assistance to the 296 local field experimental groups, and one-third are located at 29 commune and brigade level agricultural research stations where they cooperate with local technicians. In this way the KAAS is an integral part of a province-wide research, extension, and production team and all scientists at KAAS headquarters obtain practical experience on field problems. Similarly, the KAAS is made aware of field problems and has a ready mechanism for the dissemination of research results to farmers and new plant and animal strains.

There are 400 hectares of cultivated land, some 100 of which are in field test plots. There are five research institutes and an experimental farm under the administration of KAAS: The Crop Breeding Research Institute, The Institute of Soils and Fertilizer (and Crop Management), The Institute of Plant Protection, Institute for Animal Husbandry, Institute for Fruit Improvement, and the experimental farm.

The Crop Breeding Research Institute works on soybeans, maize, sorghum, millet, wheat and paddy rice and has as its objectives the development and evaluation of improved strains of these crops.

The Research Institute for Soils, Fertilizers and Crop Management has made a soils survey of the cultivated areas of Kirin province (in cooperation with local technicians). It works on soil classification and management, fertilizers, crop management, soil microbiology, and atomic energy applications.

The Institute for Plant Protection does research on diseases, insects, pesticides (including herbicides) and biological control of pests.

The Research Institute for Animal Husbandry works on management and improvement of swine, cattle, horses, sheep, and poultry, with emphasis on propagation (breeding) and feeding.

The Fruit Research Institute concentrates on the breeding and management of apples, pears, and grapes, with emphasis on adaptation and quality.

The experimental farm propagates the improved varieties of crops for use elsewhere in the province.

The KAAS participates in a research-extension-production system involving the counties, the communes, the production brigades, and the production teams. KAAS scientists (85 of them) are stationed at 29 agricultural research stations set up at the commune and production brigade levels. They assist technicians in carrying out field testing of materials from KAAS as well as selections found locally. Another 85 of the KAAS personnel, while stationed at Kung-chu-ling, assist technicians at 296 field testing sites focused

primarily at the production team level. This system provides the linkage
with production.

The KAAS accomplishments described were:

1. Survey of crops growing in Kirin province.

2. New varieties released: 5 soybean, 5 wheat, rice and millet (10
percent higher in yield than traditional varieties), 15 corn and sorghum
hybrids (corn - 30 percent higher yield than traditional varieties and sorghum
over 15 percent higher).

3. Evaluation of all new varieties at commune, brigade and production
team levels.

4. Improved cropping systems including intercropping and double
cropping.

5. Improved soil tillage techniques.

6. Surveys of occurrence of important insect and disease infestations.

7. Biological control of corn stem borer, using a wasp predator.

8. Soil survey of Kirin province.

9. Reclamation of 200,000 hectares of saline soils (out of 600,000
total).

10. New apple and pear varieties tolerant of cold temperatures.

11. Improved strains of sheep, black swine, chickens and cattle.

A visit to a plant breeding station (Big Elm Tree Brigade) of the Nan-
wai-tzu people's commune verified the general working relationships between the
KAAS and experiment stations located at the production brigade level. We met
a KAAS staff member who was assigned to and lived at the brigade level. He was
a member of the 3-in-1 combination: a scientist from KAAS, a peasant tech-
nician, and a leader concerned with production in the brigade.

Programs for breeding of corn and sorghum at this experiment station
revealed close coordination between KAAS and brigade level personnel.

Shensi Academy of Agricultural and Forestry Sciences

Formerly the Northwest Agricultural Research Institute (initiated in
1954) this organization became the Shensi branch of the Chinese Academy of
Agricultural Science in 1958. Recently its name was changed to the Shensi
Academy of Agricultural and Forestry Sciences. Its name suggests changes in
accord with national changes in the organization of agricultural research in
China.

The SAAFS is apparently organized much the same as its Kirin province counterpart. It reports to the Bureau of Agriculture and Forestry of the Shensi provincial government. Since 1970 about one-half the scientific and technical staff have spent time at the "grass roots" (brigade and production team) levels in various communes. At the present time, staff members of SAAFS are working in 75 counties at 100 locations with an average of three scientists or technicians per location. Five hundred of the 600 scientists and technicians of SAAFS have already served at the brigade or production team level.

The SAAFS has nine research units as follows: grain crops; economic crops (oil bearing, vegetables and hemp; soils and fertilizers; plant protection; animal husbandry and veterinary science; forestry; pomology; sericulture; and cotton). SAAFS scientists serve in both research and extension capacities and train technicians at the brigade and production team levels.

Nanking Agricultural Research Institute

This institute was established in 1932 and appears to be an old and respected institution. There are eight research groups: crop plants (rice, wheat, corn, sweet potato and others); industrial crops (cotton and rape); plant protection (surveillance and control); soil and fertilizers (alkaline and saline soil improvement, green manure management); animal husbandry and veterinary science (pigs and cattle); horticulture (peaches and vegetables); agricultural physical chemistry; aquatic products (breeding and disease control of fish).

There are 600 employees of whom 200 are scientists and technicians. One-third of the staff at any time work at the "grass roots" level, at three general experiment stations, 17 specific-purpose stations, and provide some contact with 100 or more brigade or team units.

We were shown a good collection of specimens of insect and disease pests of rice, corn, cotton, fruits, and vegetables. These were being used in an extension program to educate technicians and farmers to identify the pests.

The rice plots were among the best that were seen. This station has major responsibility for coordinating research on this crop in the northern region of major rice production (the Yangtze River Basin).

The research and staff at this provincial institute appeared from our brief visit to be of high quality. Discussions with their scientists suggested they are knowledgeable about their problems and are using effective means of solving them.

Shanghai Academy of Agricultural Science

An agricultural experiment station was established in Shanghai in 1956 and this academy was founded in 1960. There are five research institutes: crop breeding and culture (mostly rice, wheat and barley, cotton); soils, fertilizers and crop protection; horticulture (fruits and vegetables); animal husbandry and veterinary science; and agricultural machinery.

There are 1290 employees of whom 448 are scientific research workers. In 1974 they sent 104 of their scientists to 40 different locations to help peasants and learn from them. In turn, 12,000 peasants reportedly visited the academy during 1974.

The rice plots and field program looked good as did the rice in farmers' fields in this area. The soils work appeared to be standard with no particularly serious problems. They appear to have a viable agricultural machinery development and evaluation unit.

Kwangtung Academy of Agricultural Science

We did not visit this academy for reasons not entirely clear. It is located about 10 kilometers northeast of Canton. We were told that it administers the following institutes: soils and fertilizers; plant protection; grain crops; livestock and veterinary science; economic plants; fruits; tea (located elsewhere); and sericulture.

We visited the grain crops research institute of this academy. Apparently when KwAAS was administered under the Chinese Academy of Agricultural Sciences, this was known as the Paddy Rice Research Institute. Its mandate has been broadened to include grain crops other than rice. Since rice is the dominant grain crop, however, it continues to receive primary attention.

There are 25 people employed at the institute, one-fourth of whom are scientists. Germ plasm storage and preservation for rice is not handled here but elsewhere by the KwAAS. Except for a few of the current varieties and lines included in the breeding program, the germ plasm collection is available only through the academy proper. Apparently this academy does not serve as the germ plasm collector for indica rice in south China.

Chart 1

Organization of Science and Agriculture

Ministry of   Academy of Science
Agriculture

            Chinese Academy of
            Agricultural &
            Forestry Sciences
                                        Institute   Institute   Institute
                                            of   of   of
Atomic Energy   Other Labs   Botany   Genetics   Microbiolo
Laboratory   &
                     Institutes

            Provincial
            Governments

Provincial   Provincial
Academy of   College
Science for   of   Administrative
Agriculture   Agriculture   Authority

                                                 Consultation &
            County   Collaboration
            Organizations   (Provincial)

                                                 Consultation &
                                                 Collaboration
                                                  (National)

            Commune
            Organizations

Basic Point
Experiment   Production Brigade
Stations*   Research/Extension Unit

                  Production Team
                  Field Testing Units

*The basic point experiment stations, under the joint supervision of the
 Provincial Academy and the Production Brigade, are established at selected
 production brigades. Other brigades have research and extension units too,
 but these are not tied into the academies' basic points system.

- 11-

Chart 2

Example of Agricultural Extension System for Plant

Protection, Kirin Province, Research Network (K'o-hsueh hsi-t'ung)

Chinese Academy of Agricultural and   National
Forestry Sciences (Chung-kuo Nung-lin   (Peking)
k'o-hsueh yuan)

Kirin Provincial Academy of Agricultural   Provincial
Science (Chi-lin sheng nung-yeh k'o-hsueh
yuan)

     Plant protection section
     (Chih-pao so)

District Agriculture and Forestry Office   District
(7 such offices in Kirin) (Ti-ch'ü nung-lin-
k'o so)

     Plant protection section (Chih-pao tsu)

County Scientific Experiment Station   County*
(Hsien k'o-hsueh shih-yen chan) sometimes
called County Improved Seed Breeding Station
(Liang-chung fan-yü chan)

     Plant protection specialist
     (Chih-pao chuan-chia)

Commune Scientific Experiment Station   Commune
(Kung-she k'o-hsueh shih-yen chan)

     Plant protection specialist
     (Chih-pao chuan-chia)

Basic Points   Production Brigade Scientific Experiment   Brigade
(29 of these)   Station (Ta-tui k'o-hsueh shih-yen chan)

Production Team Scientific Experiment   Team
Small-group (Hsiao-tui k'o-hsueh shih-yen
hsiao-tsu)

Source: Interview with Hu Chi-ch'eng, in charge of Plant Protection
        Section, Kirin Provincial Academy of Agricultural Science,
        Kung-chu-ling.

*At county level and below, the research/experiment network is identical to the
administrative network. That is, the "experiment stations" at county level
and below are actually administrative organs of the state bureaucracy.

Chart 3

<u>Example of Agricultural Extension System for Plant Protection,</u>

<u>Kirin Province, Administrative Network (Cheng-fu hsi-t'ung)</u>

Kirin Province, Plant Protection Station   Provincial
(Chi-lin sheng, chih-wu pao-hu chan)

District Plant Protection Station (7)   District
(Ti-ch'ü chih-pao chan)

County Scientific Experiment Station   County*
(Hsien k'o-hsueh shih-yen chan)

Plant-protection specialist

Source: Interview with Hu Chi-ch'eng, in charge of Plant Protection Section,
         Kirin Provincial Academy of Agricultural Science, Kung-chu-ling.

*From county (hsien) level downward, the administrative network is identical
to the research/experiment network.

Higher Education in the Agricultural Sciences

Following the Cultural Revolution in the late sixties, major and far-reaching changes have occurred in all colleges and universities of the People's Republic of China. The changes have had special impact on the agricultural colleges since with few exceptions these institutions have been moved from urban centers of population to rural areas. In many instances the massive task of building new campuses with attendant facilities is still not complete. Curricula, administrative patterns, instruction and evaluation procedures, course schedules, and admission requirements have undergone revision during this period. It is essential to recognize that the process of evaluation and reorganization of these institutions is still under way although certain new formats are emerging. Although many colleges are back in operation, a few have not initiated full schedules of class work because the moves have not been completed and student housing is lacking.

During the Cultural Revolution professors at agricultural colleges were assigned to work at various communes or state farms. Following this period, which may have varied from one to two years, professors have been involved revising courses and curricula as well as preparing new course material to adjust to the decrease in the amount of time that is allotted for formal college training (i.e., from 4 to 3 years at most colleges). The textbooks formerly available are no longer considered appropriate for the new teaching format and are used only as reference sources. Students may be provided with new materials prepared by teachers; these may be in the form of supplements to or copies of lecture notes or in some cases complete manuals based on revised material. New, completely revised texts are in the process of preparation, and will soon be available at most institutions. The magnitude of effort involved in this aspect of the changed educational process is worthy of note.

The patterns of evaluation of student performance may vary widely depending on the specific decisions made by a college or even independently by a department at a given college. These may range from written examinations to group discussions and oral quizzes.

With few exceptions, students at agricultural colleges come from communes in the province in which the college is located. Candidates are chosen by a selection committee of the commune. The primary criteria for selection include: willingness to make a commitment of service to their fellow workers, political attitude, moral character, health status and recommendations of fellow workers and teachers. Students also must have an acceptable scholastic record in middle school. The schooling prior to college may vary among provinces with arrangements of five years in primary and four years in middle school being typical. However, many students leave for communes after two years of middle school. Students with this background are provided pre-college course training to bring them to the educational level of students with three years at the middle school.

There are advantages in the new plan since most students entering college will have had at least 2-3 years of practical experience on the farm. This provides a basis for instruction that may result in more rapid progress in many instances. The age of most entering students is about 20.

When students arrive at college they have either been assigned or have selected a specific curriculum. In addition to the program within the area of their specialty, they normally take courses in botany, zoology and chemistry. Agronomy and horticulture students will usually be required to take courses in plant protection such as entomology and plant pathology. Mathematics above that presented at the middle school may not be required of general agriculture majors.

Most agricultural colleges currently require training in one foreign language; the language usually will be English or Russian. Apparently most students are electing English at this time. Some students will have had some English training in middle or primary school.

Students are actively involved with their professors in the teaching process. They may work with their professors in developing the programs that will be followed. Approximately one-third of the time of each student is spent in classes under field conditions in selected communes so that course work can be seen in application. This provides the basis for learning theory in practice; usually the trips to the communes by teachers and students provide opportunity not only for more effective teaching but also develop a teacher-student relationship which is designed to enhance the learning process.

Students in animal husbandry may work in a brigade at a commune that specializes in production of swine or other animals or they may be assigned to a veterinary clinic.

It is expected that students who are having difficulty in a given course will be helped by fellow students and their teachers. It is considered important to develop a cooperative spirit of mutual assistance in order to master the subject material. There is a desire to minimize competition for top grades. There could be some student failures under this system, but this would be rare. Students in certain departments may be asked to complete a specific research project on a given topic. They would be expected to undertake observations and research under field or applied conditions.

Students are also actively involved in evaluation of their teachers and fellow students. In other words, they are expected to criticize each other in a constructive way. The format for this evaluation may involve a formal session once a month in which the class as a whole meets with the teacher to discuss ways in which instruction might be improved. Or, students may schedule separate discussions. Apparently this procedure of student evaluation as well as other involvement by students in university governance is quite a change from the traditions of the past when the professor was in complete and authoritative control of all phases of the education process. The close personal relationships that develop between teachers and students probably make this interaction possible. The main concept is to avoid critique on a personal basis.

Graduate study as it is currently organized in the United States has no equivalent counterpart in the Chinese agricultural colleges. However, individuals selected to be teaching assistants have many experiences in common with American graduate students. The assistants help teach courses, prepare materials for class, and present laboratory lectures. They usually work under the supervision of a single senior professor. He may ask them to practice lectures with him prior to presentation or he will sit in on their laboratory or lecture presentations and then advise them how to improve. Assistants also help with field and laboratory projects since professors at most agricultural colleges do not have the equivalent of civil service technical aides or secretarial help. Assistants will frequently be asked to initiate a specific research project under supervision of the professor who will assign readings in the literature that are relevant to the project. In this way, as an assistant makes progress he will complete many phases of the advanced training that American students receive except that the Chinese students will be expected to carry a relatively full work schedule. They are, unlike graduate students in the United States, considered members of the faculty.

At present there is a moratorium on the appointment of new professors. The decisions have not been made as to the exact titles for equivalent positions and the procedures for selection and appointment. Salary advancements within lower ranks have been made but there have been few if any advancements in title on an official basis.

Professors may continue to work throughout their normal life span. As yet there is no designated retirement age for college professors. Faculty members currently holding the title of "Professor" will continue to receive full pay as long as they live.

Professors and administrators below age 55 (unless excused for physical disability or other reasons) are expected to spend about 30 days per year in some type of physical labor or public service work. This usually presents no problem for faculty in agricultural science because their field activities and work with research projects make it possible to complete this assignment without need for extra work assignments.

At some colleges administrative heads have been appointed; at most, individuals are serving on an "acting" basis.

Publications

There is a trend at present for collective authorship of research publications. The pressure for individual recognition which may have existed in the past is now less important. There is probably a greater tendency for results of research to be shared at a preliminary stage in meetings and in discussions with other workers interested in the same problem.

Most journals of agricultural science have been discontinued and the exact format of new journals has not yet been determined. Results of most research projects reported to us had not been published. However, many projects have only recently been initiated and results have not yet been confirmed by repetition of the experiments.

Organization of an Agricultural College

The Northwest College of Agriculture at Wu-kung in Shensi Province was founded in 1934 and has just completed four decades of growth. Most expansion has occurred since 1949. The area of land has increased from 28 to 123 hectares, and the number of laboratories for research and teaching has increased from three to 84. After the Cultural Revolution there was a major change in teaching and research activities. Prior to this change, students followed a traditional program of four years of study. This has now been reduced to two years for agronomy or three years for veterinary science, based on the assumption that all new students will have spent at least two years on a commune. Students may enroll in one of eight departments: agronomy; forestry; plant protection; horticulture; animal husbandry and/or veterinary science; agricultural engineering; agricultural economics; and hydrology.

The library of this college has over 410,000 volumes at present. A number of key journals in English are currently received. Large reading rooms are available for each of the special subject areas, and key reference texts are held on reserve in these rooms.

Students may follow study programs of specialization, including agricultural crops (agronomy), fruit crops, animal husbandry, plant protection, irrigation, water conservation, design and manufacture of agricultural machinery, use and maintenance of agricultural machinery, forestry, and veterinary science.

Before the Cultural Revolution there were 4000 to 5000 students. However, both students and faculty were sent to various communes and all classes were suspended for about two or three years.

At present there are 1476 staff and teachers, including 370 cadre (administrative personnel), 63 professors and associate professors, 221 lecturers (instructors) and 228 assistants. There are 1200 students. In addition to faculty at the college, various scientists at the nearby academy are invited to present lectures. Also, scientists and professors from other colleges offer lectures and seminars.

As is true for all other organizations, the governing body of the college is called the "revolutionary committee." This committee usually includes the president or director, certain administrative officers, faculty members, staff, and students. There are usually 10 to 12 members on the committee. Decisions as to policy matters are subject to final approval by higher administrative authorities at the provincial level.

THE STATUS OF AGRICULTURE

From the beginning of recorded history agriculture has played a leading role in the culture, life style and economy of China. Even today between 80 and 85 percent of the total population is engaged in agricultural activity of one kind or another.

Land Area and Land Use

The People's Republic of China (PRC) has a total land area that is somewhat larger than the original 48 states of continental U.S.A. as indicated in table 1. Although the total land mass of the PRC is 973 million hectares, compared to 768 million in the U.S.A., much of the land area of China is unsuitable for agriculture, as it is too mountainous, or too dry. China's arable land area is estimated to be only two-thirds that of the United States. However, China's land is cultivated intensively by the employment of double or triple cropping wherever climate permits. Consequently, the sown area in the PRC was estimated in 1964 to be 150 million hectares, compared to a sown area of 116 million hectares in the U.S.A., where much arable land was idle because of production or area restrictions. Were this comparison made for 1974 when much of the idle land in the U.S.A. has been returned to production, the figures for sown land would probably be 155 million hectares in the PRC and 136 million in the U.S.A. China continues a major effort to further increase double and triple cropping wherever possible and thereby to increase production to meet the food needs of its large and rapidly increasing population. China's population is now estimated at 850 to 900 million and is probably adding between 17 and 18 million people per annum.

A further analysis of China's land use (table 2) indicates that 15.3 percent of the total land area is classed as arable while 17.8 percent is of little or no value for agriculture, animal industry or forestry because it is too arid, at too high an elevation, or devoid of tillable soil.

Grassland, mostly in semi-arid areas, constitutes 27.8 percent of the surface, and forested area, about 7.9 percent. Another 31.3 percent of the total area has been denuded of forests, cultivated, and abandoned over the centuries but is suitable for reforestation. Perhaps much of this area, even if reforested, would be incapable of producing commercial forest products though reforestation would eventually reduce erosion, runoff, and silting, and also stabilize stream and river flow.

Despite the discrepancies and inadequacies of data on land use, it seems clear that China's effort to increase food production during the next decade must concentrate on raising production on areas already under cultivation by increasing yields per hectare and further multiplying cropping intensity (e.g., double and triple cropping) wherever possible. Development projects will be costly and can at best expand the cultivated area only modestly and slowly during this period.

Farm Organization

The PRC's present agricultural production organization is the result of the evolution or changes that have occurred since "liberation" of the peasant masses in 1949. (Peasants constitute 80 to 85 percent of the population.)

Centuries of peasant abuse by landlords, tax collectors, warlords, and moneylenders, set the stage for the popular uprising in support of the communist armies in the late 1920's, through the 1940's. It was the promise of freedom from these abuses and the redistribution of land (which was actively implemented as soon as a "soviet army" took over an area in the 1930's) that permitted the establishment of the Communist government in 1949.

One of the first major undertakings of the new government of the People's Republic of China was land reform. This reform has passed through several phases.

Phase One: Many million hectares of land were distributed by Chinese Communist groups before they established the PRC in 1949. At that time the program was expanded nationwide. In this stage (1949 to 1953) ownership and management was left to the individual farmer.

Phase Two: In 1953 a program of agricultural socialization began. Although land ownership and management was left to the farmer, groups of neighbors were encouraged to form "mutual aid teams" for pooling their efforts and resources, and increasing production.

Phase Three: Formation of Agricultural Producers Cooperatives. The government (Party) encouraged and forced the amalgamation of neighboring "mutual aid teams," formed in Phase Two, into agricultural cooperatives. Each household gave land and capital to the cooperative and in return received remuneration proportionate to the land shares and labor contributed to production. These cooperatives were controlled by a committee, including a chairman and an accountant.

Phase Four: Collective Farms. Later the Party combined a number of the semi-socialist Agricultural Producers Cooperatives into collective farms, generally consisting of 100 or more households. As this program evolved the collective farms often grew larger. They are now known as communes.

Today communes cultivate 90 percent of the arable land. The average size of a commune, though it varies considerably from place to place, is 1900 hectares. **They range in size from 2000 to 20,000 families and 8000 to 80,000 people: one near Canton, Shi Ch'iao, had 19,000 households and** 74,000 people. In 1973 there were 50,000 communes compared to 74,000 in 1964. Consequently the number of production brigades has increased to improve efficiency.

Communes are organized into production brigades and production teams. The production team is the key organizational and accounting unit. It is at this level that production management decisions are made, such as what crops are to be grown, how they are to be cultivated, what inputs will be used, etc. There are conflicting reports on how many production teams are in operation, with estimates ranging from four to eight million. The average production team is comprised of 20 to 40 families with 88 to 176 people. They are responsible for the cultivation of from 8 to 38 hectares of cropland.

An incentive remuneration system has been built into the commune system at the production team level. This is called the Labor Day Work Payment System whereby farmers (families) are given larger or smaller shares of the collective net income, depending upon the amount of labor (based on work points earned and tasks assigned) that they contribute during the year. Consequently, it does not mean that doing a day's work ensures that a "labor day" will be credited.

Phase Five: State Farms. Apparently an ultimate goal of the government has been to convert many of the present communes to state farms. In 1965 there were 2000 state farms compared to only 18 in 1949. State farms in 1965 were estimated to constitute less than 4 percent of the total cultivated land area. Many of them are located in the flat plains of the three northeastern provinces where farm mechanization is well advanced. The number of state farms is increasing but no recent estimates on numbers were found. Whether the conversion of communes to state farms will continue is uncertain. Apparently there is growing satisfaction with performance of communes, and reportedly some reluctance to go on with Phase Five.

Private Agricultural Land

Private land constitutes less than five percent of the cultivated land area of China. Most of this is in the form of very small family plots that supplement food production and add to family income. Much of this production, especially near large cities, is devoted to high income crops that can be sold in the city markets. The original size of the plot or number of family plots cannot be expanded even if families increase in size; this is one incentive to keep families small.

A small number of private farms exist in frontier areas where the population is sparse and dispersed and does not justify the establishment of communes.

China's Agricultural Production and Potential

Much of China's total agricultural production is concentrated in drainage areas and deltas of the three great river systems, namely, the Huang Ho (Yellow), the Yangtze, the Hsi Chiang (Pearl), and their tributaries. The "Manchurian" Plain, comprising the provinces of Heilungkiang, Kirin and Liaoning in the northeast, is also of great importance in food production. Virtually all of the food and fiber production of the nation is concentrated in the eastern half of the country, the majority of it within the eastern one-third between 21 and 48° N. latitudes. This range in latitudes encompasses a diversity in climatic zones and crops as great as those between Miami and northern Minnesota. Western China (Tibet) is largely mountainous, too high or arid for agriculture and suited only for grazing (Sinkiang-Inner Mongolia).

China has a larger irrigated land area than any other country in the world. Slightly less than one-third of the cultivated crop land or approximately

33.5 million hectares are irrigated, compared with roughly 15.8 million
hectares in the United States.

During the past several years large work forces have been engaged
in expanding or improving the irrigated areas by digging new canals, constructing minor water storage ponds, or reservoirs, and installing wells and
pumps. In some areas where lack of drainage has been a problem, new drainage
systems have been dug. All of these improvements add to the potential and
stability of food production. There is every indication that these types of
winter work projects will be continued and expanded in the years ahead, especially along the Yellow River systems where the construction of large dams
and reservoirs has been handicapped and made only partially effective because
of rapid silting.

The Present Status of Crop Production

The PRC government has not published a detailed description of crop
production statistics since 1960. Thus the reliability of current production
estimates by foreign institutions is doubtful. Apparently most have been based
on projections from 1960 data supplemented by fragmented data on production increases by various Chinese government agencies. A high experimental error should
be expected. The U.S. Plant Studies Group was unable to obtain any new information on crop production statistics during the course of the trip. Data presented
in table 3, giving acreage and production estimates by Stavis, FAO, and the
USDA reveal considerable variation for different crops as well as for total of
all crops. The estimates of the USDA seem to be considerably lower than those
of either Stavis or FAO. The total production estimates for all crops in the
1971 record crop season are 247 and 249 million metric tons by Stavis and FAO,
respectively, whereas the USDA estimate was reported as 228 million tons, considerably lower than the PRC government announcement of a record 250 million tons.
Although the Stavis and FAO estimates of total production are quite similar, the
estimates for specific components are very different. Estimates by Stavis for
rice and wheat exceed those of FAO by 5 and 10 million tons respectively. The
Stavis estimate for wheat production appears unrealistically high. On the
other hand, the FAO estimate for tuber production is 10 million tons higher than
that of Stavis. These variations in estimates for different crops, however, tend
to be cancelled out and the resulting estimates of total production are very
similar.

Regardless of details in estimates, agricultural production has certainly increased greatly since 1949. Much of this increase apparently took place between 1957 and 1971 when production of grain rose by some 60 to 65 million tons. There is reason to believe that the modern sector, constituting only 20 percent of the cultivated area, accounted for 42 percent of this production gain.

The government of the PRC reported that it had reached grain production self-sufficiency in 1971, when total production was reported to have reached 250 million metric tons. However, as a result of the severe drought in 1972, grain production was estimated to have dropped about 10 million metric tons. The result was that during the 1973-74 crop reporting season (of the USDA), China was estimated to have imported more than 6 million metric tons of wheat and 2 million tons of corn. After seeing the good-to-excellent crops that are now growing everywhere we traveled, it would appear that the current imports are probably being used to replenish grain reserves.

Whatever the difficulties in nationalizing production statistics, the best criteria of progress in agricultural production are the physical condition and vigor of the people and the observed conditions of the crops.

Throughout the course of this trip, we saw no evidence of hunger or food shortages among the masses, either urban or rural. Moreover, people everywhere appeared to be adequately fed and physically vigorous and energetic. All were adequately and comfortably clothed. The standard of living - especially as it refers to nutrition, clothing and medical care - of the masses in the People's Republic of China is far better than that of the masses in most Third World nations of Asia, Africa and Latin America. This is a good measure of progress in agricultural production.

Recognizing this great improvement in agricultural production, the question arises: How did this come about? Many factors are obviously involved and many changes must have contributed to the improvement, including those in the social, economic, and political structure of the nation. The feudalistic system of pre-1949 was replaced by a system in which production incentives have been incorporated. Consequently the masses who work the land on communes are stimulated to produce more in anticipation of increased income.

Within the past decade a major effort has been made by the government to decentralize the administrative bureaucracy and make it more responsive to the needs of the rural people. A "back to the grass roots" movement has also been implemented to make scientists, including all who work on different aspects of agricultural sciences, more responsive to production problems.

Scientists have introduced new technology into production at the grass roots level. This new technology includes the development of high-yielding, disease-resistant crop varieties. The benefits of the use of these varieties has been widely demonstrated and "popularized"; the development and introduction of improved cultural practices, e.g., proper fertilization including the correct use of chemical fertilizers, rates and dates of sowing, better weed and pest control; the development of effective intercropping and multicropping systems to increase cropping intensity and total production; the encouragement of better water utilization and improvement and expansion of irrigated areas; and the government's expansion of the production capacity of vital inputs such as chemical fertilizers, pesticides, pumps, machinery, transportation, and storage facilities.

It is not necessary to indicate how all of these interrelated factors that affect production have been manipulated in order to appreciate some of the recent improvements in production. Mention will be made of only a few of the key factors that have been modified. This includes the introduction of improved varieties, expansion of production, introduction and use of chemical fertilizer, improved pest control, increased cropping intensity, increase in price paid to farmers for products and decrease in price of production inputs such as chemical fertilizers.

The Introduction and Widespread Use of Improved Crop Varieties

The first high-yielding dwarf indica varieties of rice were introduced into commercial production in south China in the early 1960's; similar types of japonica were already in use in the north. The indicas were developed in the PRC by Chinese scientists from breeding programs initiated in 1956. The first seed of IR 8 was introduced into China in late 1967. These Chinese indica varieties had high genetic yield potential, were fairly responsive to high levels of fertility, and relatively resistant to lodging and to diseases. They spread rapidly. By 1965 the southern area sown to high-yielding rice varieties was estimated to be 3.3 million hectares. By 1973 these varieties reportedly were being grown on 6.7 million hectares. Apparently the area is continuing to expand for we saw very few tall-strawed rice varieties during our travel; those we did see were at experiment stations.

Similarly, high-yielding winter wheat varieties have been developed since the early 1960's. The first improved wheat varieties of that period appear to have been tall-strawed, disease-resistant types with considerably higher yield than the older ones. These improved varieties were being grown in 2.5 million hectares by 1965 and must cover much larger areas at present. It seems that the first high-yielding dwarf winter wheat varieties were increased in 1972. A variety Tung-fang-hung 3, bred at Peking, appears to have about twice the yield potential of the earlier improved tall varieties. During 1973 multiplication of another high-yielding dwarf winter wheat variety called I Fung 3 was begun by the Northwest Agricultural College at Wu-kung, Shensi Province. There may also be other new dwarf varieties in existence. The dwarfing genes for these new dwarf wheat varieties were obtained from inter-crossing Korean dwarf winter wheats (Seouwan 86) with the best tall Chinese varieties. During 1973 and 1974, 5000 and 15,000 tons, respectively, of seed from several varieties of dwarf Mexican spring wheat varieties were purchased and introduced.

The use of hybrid corn and sorghum has contributed significantly to increasing yields of these crops, especially within the past five to seven years. In some areas such as Kirin Province in the northeast, 60 to 70 percent of the area sown to corn is said to be in hybrids. In other areas visited, the percentage of area sown to hybrids is much less. Sorghum hybrids are also gaining in popularity. Improved varieties of millets, soybeans, barley and many fruit and vegetable crops have been developed and are more widely grown. All of this work has made substantial contributions to raising yields and production during the past seven or eight years. Continuing benefit and yield increases will be forthcoming from the more widespread distribution and use of the currently high-yielding varieties, and other even better varieties that should be produced within the next five to ten years.

The Improvement of Cultural Practices, Including the Use of More Chemical Fertilizers

China has long been known for its effective use of decomposed organic matter to maintain a fairly high level of soil fertility. Had it not been for the skillful widespread use of this practice over past decades, crop yields in China would have fallen to a much lower level than they did (as was the case in India, Pakistan and many other countries in Asia, Africa and Latin America) before the advent of the use of chemical fertilizers for basic food crops. Over many decades, and even centuries, crop wastes including rice and wheat straw, sorghum and maize stalks, weeds, leaves, and rice glumes have been combined with human and animal waste in compost pits or piles to be fermented and decomposed before being returned to the fields to maintain soil fertility. When one compares the basic level of soil fertility of cropland in China with that of cropland in India, Pakistan or many other developing nations in Asia, Africa or Latin America where the organic matter is generally used as fuel, the beneficial effects of the Chinese practice are very apparent. The value of organic fertilizers is universally appreciated in China. In all probability one of the principal reasons for encouraging hog production in the past two decades - in addition to providing meat and an important source of income from its sale by family producers - is for the production of manure, essential to effective composting of other plant wastes. The Chinese have learned the art of effectively handling enormous quantities of compost without endangering human health. In fact one seldom sees a fly in China despite the enormous quantities of compost, due to the skillful use of insecticides around the fermentation pits and piles. Both chlorinated hydrocarbons and organic phosphates have been used for this purpose.

Virtually no chemical fertilizers were used in China before 1960 and soil fertility was maintained to the extent possible by the use of organic manures, green manures and silt and mud from rivers, canals and lakes. Although there has been a rapid expansion in chemical fertilizer production in the past seven or eight years, it is difficult to estimate the amount of plant nutrient produced. In the past the PRC government has reported production in tons of fertilizer rather than tons of plant nutrients. Since a considerable number of different kinds of fertilizers with different levels of nutrients are produced, this greatly complicated interpretation of production figures. The data in tables 4, 5 and 6, nevertheless, clearly indicate a rapid increase in chemical fertilizer production and use.

The consumption of chemical nutrient nitrogen in 1973 was estimated to be 4.2 million metric tons, of which 2.4 million tons were produced domestically and 1.8 million tons were imported. For the past several years the PRC has been the world's largest importer of nitrogenous fertilizers.

The percentage of nutrient nitrogen coming from chemical fertilizer has increased spectacularly in recent years. Table 5 shows the high and low estimates for plant nutrients applied from both organic and chemical fertilizer sources in 1972. Using the high-level estimate, it is considered probable that 14.8 million tons of nutrients would be furnished from traditional organic fertilizers and 4.5 million tons from chemical fertilizer. In this case the percentage of nutrients derived from chemical fertilizers is 23 percent. If the low estimate is used, 6.0 million tons and 3.6 million tons of nutrient would be from the organic and chemical sources, respectively. In this case the percentage of plant nutrients derived from chemical fertilizer would be 37 percent of the total.

One of the greatest contributors to increased domestic fertilizer production since 1964 has been the installation of many relatively small, widely dispersed nitrogen factories producing anhydrous ammonia and ammonium bicarbonate from lignite (table 6). We had an opportunity to visit one of these factories near Shanghai.

This small fertilizer plant is one of many said to be located around the countryside in China. It is unique in that it produces ammonium bicarbonate or NH_4HCO_3 (17% N), a product that is essentially unknown as a fertilizer in the United States.

This plant was first set up as a Shanghai county-operated unit in 1959. Its capacity at that time was only 800 tons of NH_3 per year. The plant was later renovated and enlarged and now produces 20,000 tons of NH_3 annually, of which 80 percent is converted to solid NH_4HCO_3 and the remainder is distributed as aqua ammonia, or NH_4OH (42.5% N). The ammonium bicarbonate now sells for 140 Yuan/metric ton ($70/MT) and the aqua ammonia for 75 Yuan/MT ($37.50/MT). This is equivalent to $.187 per pound of N (nitrogen) in the NH_4HCO_3 and $.136 per pound of N in the NH_4OH. This is very high by July 1972 U.S. prices of $.045 and $.017 per pound of N from urea and anhydrous ammonia, respectively, but much cheaper than July 1974 U.S. prices of $.30 and $.17, respectively, for these same products. Prices are held down because fertilizer is usually used locally and any transportation is by boat as the plant is located on a navigable canal.

The source of raw material for this plant is lignite, brought to Shanghai from the interior (Shensi Province). The lignite (screened and sized to make briquettes if it is too fine) is reacted with steam in a furnace, and the gases N_2, H_2, CO_2, CO, O_2 and CH_4 are driven off. The CH_4 is used to make

methyl alcohol (CH_3OH). Oxygen is removed with copper. The CO_2 and CO
are separated out by compressing the remaining mixture to 100 Kg/cm^2, leaving
N_2 and H_2 gases. These are then placed under 200 to 300 Kg/cm^2 pressure
and just under $500^{\circ}C$ temperature and NH_3 gas if formed by the Haber process.
Thus, the NH_3 is synthetic and not the typical by-product gas from the manufacture of coke.

NH_4HCO_3 is produced by passing CO_2 through an NH_4OH solution. A
fine white product results which is said to cake if allowed to sit too long.
There was no evidence of storage at this location and the product appeared
to be moving out to the farms very rapidly.

The aqua ammonia is said to be pumped into the irrigation water for
rice and in some cases wheat or barley. This product is used more during
the rush season when it can be applied directly to the field. It is trans-
ported from the plant in concrete tanks kept in boats.

Workers for other plants have been trained at this location - a
total of 4000 since 1959.

Fertilizer usage in this county reportedly has increased from 37
to 50 Kg of N/hectare in 1959 to 195 to 250 Kg N/hectare today.

During the past year the PRC has signed contracts for the installation of ten 1000-ton-per-day anhydrous ammonia factories together with the
urea plants necessary to convert most of this anhydrous ammonia to solid
fertilizer. The operation of these plants in 1977 and 1978 will add approximately 2.7 million tons of nutrient nitrogen production. The worsening world-
wide shortage of fertilizers and soaring prices together with the need to
assure an adequate supply to meet rapidly growing domestic demand undoubtedly
contributed to the government's decision to make this huge investment in
nitrogenous fertilizer production. The capital investment in the factories,
pipelines, warehouses, railroad cars and such, will probably run between 800
million and one billion dollars. China has the raw materials, either natural
gas or crude oil, to produce all the nitrogenous fertilizer it will need and
thereby protect itself from the soaring prices of fertilizer on the international market.

This great expansion in nitrogenous fertilizer production will certainly result in further additions in grain production. Although the growing
crops we saw everywhere generally looked good, many of them showed some signs

of nitrogen deficiency. This was true of rice, corn, sorghum, millet and cotton. Rice generally seemed to have been fertilized better than other crops.

The use of phosphate fertilizer has been increasing rapidly during the past seven to eight years, especially for many crops that are grown on the red soils of the southern provinces. Presently, an approximate one-third of the nutrient phosphate (P_2O_5) is imported. Phosphate rock is also being imported in substantial quantities.

Increasing Cropping Intensity

Crop intensity has been enhanced over the past few years - especially in areas where crops can be grown throughout the entire year or most of the year. Where this is possible, the land is sustaining growing crops nearly every day of the year. In many areas three crops are now produced per year where only two were previously grown, e.g., rice-rice-wheat, or rice-rice-barley. In other areas two crops are produced where one was grown before, e.g., wheat-corn, wheat-sorghum, or wheat-cotton. The intensity of cropping has been made possible by the development and use of high-yielding, quick-maturing varieties combined with the increased use of chemical fertilizer.

Many combinations of intercropping and mixed cropping are under evaluation. Plant population densities are being increased in combination with higher levels of fertilization and the use of varieties with greater resistance to lodging. All of these practices are being evaluated currently under semi-commercial conditions. The best combinations of cultural practices will undoubtedly be used widely in the next few years to further increase production.

Expanding and Improving Irrigation

In recent years increased attention has been given to expansion and improvement of irrigation systems, especially during the winter season. Many new diversion canals, catchment basins, and wells with mechanical pumps have been installed. Although the delegation was given no data on the area that has been provided supplementary or full irrigation as a protection against drought, one estimate is that more than 5 million hectares have benefited from such projects since 1965. Much of the improvement in irrigation systems is made during the winter when there is less demand for field labor.

The Outlook for Expanding Crop Production During the Next Decade

There is little doubt that China's crop production can be increased substantially during the next decade. Whether specific crop increases will be

of the order of 20 percent or 50 percent will probably depend on the crop
and its level of yield at the present time. The level of rice yields, based
on our observations in many places, is now relatively high, and consequently
it will be much more difficult to increase the per hectare yield of rice by
50 percent than of corn, or sorghum. The data presented in tables 7 and 8 indicate that the 1967-71 five-year average per hectare yield for the major
crops in China is only 50 percent or less - with the exception of rice - than
that of the United States. These data again raise the question of the reliability of the statistics.

If the data on comparative yields for each of the major crops in China
versus those of the United States are reasonably correct, then China's yield
can be greatly increased in the next decade. This is especially true as more
than one-third of the sown area in China is irrigated.

It is probable that we have been privileged to see China's agriculture in an unusually favorable year. There is no doubt that crops are
good-to-excellent. It is entirely possible that 1974 harvests will set a new
record.

Animal Production

Although our delegation was officially concerned only with plant
sciences and crop production it was impossible to ignore livestock production,
especially hog production which is such an integral part of the cropping system.
The estimated livestock numbers for the PRC are given in table 9.

Virtually all of the sheep and goats are raised in the grasslands
of Sinkiang and Inner Mongolia and the more mountainous areas of neighboring
western provinces. Many of the cattle can be found in these same areas. Of
the other larger animals, most of the buffalo, donkeys, and mules and many of
the horses are present in the agricultural areas of eastern and northern China.

Undoubtedly, the most important animal in the economy is the hog; it
is estimated that in 1972 there were 250-260 million in China, more than a
400 percent increase over the estimated 1949 population. This is approximately
four times greater than in the United States, which prides itself on its pork
production.

Hogs are produced very differently in the two countries. In the
United States the objective is to produce a 100 kilo pig in the shortest possible

number of days, nearly always under six months. To do this it is necessary to use balanced rations, based largely on corn and soybean with the addition of the necessary minerals, vitamins and antibiotics. In China the pigs are generally sold at eight months to one year of age when they will generally weigh 70 to 100 kilos. The minimum weight of pigs purchased by the state is 50 kilos.

With the exception of the northeastern provinces where winters are both long and severe and where it is therefore necessary to feed pigs considerable amounts of grain (mostly corn), the pig is fed waste materials not suitable for human food. Depending upon the area of the country and season, their diets will consist of wastes from vegetable production, including leaves of cabbage, corn husks, sweet potatoes (both leaves and tubers), many kinds of weeds and water plants (especially Alternanthera, Pistia, and water hyacinth), **the leaves and** dried pods of soybeans, peanuts and rape. Included in their feed also, when available, is the bran from rice, wheat, sorghum and millet, together with small amounts of cottonseed cake, soybean cake, linseed- and peanut-meal and even rice hulls. These ingredients will vary depending upon local availability, and the season of the year. Since many of the ingredients are highly fibrous and indigestible, they are usually ground, soaked, and fermented for varying lengths of time before feeding.

Apparently there has as yet been no research in China to improve the essential amino acid balance and nutritional value of cereals. If this were done, it would have a direct effect on improving the nutritional value of cereals for humans. It would also be of great importance to increasing livestock production, especially hogs.

Genes are now known that will improve the essential amino acid balance and nutritional value of corn, sorghum, and barley - all of which are used both for human food and livestock feed. This aspect of cereal improvement merits considerable emphasis in China.

The pigs raised in China, unlike those in the United States, do not compete to any appreciable degree for grain that is potential food for people.

Most pigs grown in China are produced by individual rural households; each family generally raising one or two. The commune usually provides piglets as it maintains the breeding stock. The pigs are not only a source of meat but add to the individual family incomes. Perhaps above all else the pig is valued by rural people for the manure it produces, which is one of the key ingredients for composting plant wastes into organic manures.

Population and Food Production

Can the People's Republic of China increase food production at a rate that will continue to improve the standard of living of its large and growing population? Enormous progress has been made by the PRC in the 25 years since liberation. An agricultural production revolution has removed the threat of famine and provided adequate food for the people. This is a tremendous agricultural achievement. It is especially praiseworthy when one considers the population size. Nevertheless, this size and growth rate constitutes a tremendous and ever-present threat to the nation's ability to further improve its standard of living.

The last complete census was made in 1953 when the Chinese population was officially reported to be 583 million. Several different estimates indicate that it now has a population between 870 and 900 million. There is some indirect evidence that the population will pass the 900 million mark before the end of 1974.

The PRC officially proclaimed that it had achieved self-sufficiency in food production in 1971 when it produced 250 million metric tons of rice, wheat, miscellaneous grains, and tubers. At that time China's population was probably around 855 million. If self-sufficiency has been achieved, the per capita consumption of these basic foods should be about 0.293 metric tons or 293 kilograms per person per year. Estimating the present population at 900 million and an annual population growth rate of 2 percent, population is increasing at about 18 million people annually. Assuming the same per capita **requirement** of 293 kilograms, this current 18 million annual increase will require an increase in food production of about 5 1/4 million metric tons annually simply to maintain the 1971 standard of diet.

China has the agricultural production potential to meet its food needs if a coordinated and aggressive effort is sustained; even so, supplying adequate food needs will become increasingly more difficult unless population growth is greatly reduced. The PRC is concerned about population growth, and has the political, administrative, educational, and disciplinary structure to deal with the problem at all levels. Within the past year the government has begun to implement a more forceful program to slow population growth, in an effort to avoid the food shortage crisis it might otherwise face.

Table 1. Land area of the United States and the People's Republic of China[1]

Classification	United States Million Hectares[2]	People's Republic of China Million Hectares
Total land mass[2]	768	973
Arable land[3]	156	107
Sown area[3]	116	150[4]

[1] Source: Agriculture in the United States and the People's Republic of China 1967-71, Foreign Agricultural Economic Report 94, Economic Research Service, USDA, February 1974.

[2] Data represent the 1964 areas of the 48 contiguous states, and proportions in various categories in China over the past decade.

[3] Arable land includes cropland used for crops, soil improvement crops, or land that is idle; it excludes pastures in both countries. Sown area is cultivated land less fallow land.

[4] Sown area in China is 40 percent more than arable area because of multiple cropping.

- 34 -

Table 2. Land use in the People's Republic of China, based on 1963 estimates[1]

Use Pattern	Area: Million Hectares	Percentage of Total Land Area
1. Total arable land	147.4	15.28%
Cultivated	(110.6)	(11.46%)
Uncultivated	(36.8)	(3.82%)
2. Forest area	76.4	7.92%
3. Land suitable for reforestation	301.5	31.25%
4. Grassland	268.0	27.78%
5. Other - deserts, high mountains, etc.	171.5	17.77%
Total land area	964.8	100.00%

[1]Source: Praeger Special Studies in International Economics and Development (Chao Shih-ying), London, 1972.

Table 3. Three different estimates of crop production in the People's Republic of China covering the period 1957-1973

Crop	Estimates of area sown (million hectares)			Estimated production in millions of metric tons						
	Stavis		FAO[1]	Stavis		FAO[1]	USDA[2]			
	1957	1971	1971	1957	1971	1971	1970	1971	1972	1973
Rice[3]	32.2	34.2	33.8	87	111	106	100	103	98	103
Wheat	27.5	27.5	28.5	24	42	32	27	26	28	28
Miscella- neous Grains[4]	50.6	58.6	56.8	53	70	75	73	72	65	72
Tubers	10.5	10.5	15.7	22	24	36	25	24	24	25
Total	120.8	130.8	134.6	186	247	249	225	225	215	228*

[1]Source: "Making Green Revolution - The Politics of Agricultural Development in China," by Benedict Stavis, Rural Development Monograph #1, Cornell University, 1974.

[2]Source: "The Agricultural Situation in the People's Republic of China," Economic Research Service, ERS - Foreign 362, U.S. Department of Agriculture, May 1974.

[3]Rice production figures are for unmilled paddy.

[4]Miscellaneous grains include barley, buckwheat, corn, millet, oats, pulses, rye and sorghum.

[5]Tubers are mainly sweet potatoes and Irish potatoes converted to grain equivalents at one-fourth of fresh weight.

*Ting Chung of the Ministry of Agriculture, People's Republic of China, estimated in October 1973 that the total 1973 grain production would exceed the 1971 record total of 250 million metric tons.

- 36 -

Table 4. Estimated production, imports and consumption of chemical nitrogenous and phosphatic fertilizer in the People's Republic of China, expressed in thousands of tons of nutrients, 1960-1973[1]

Year	Production		Imports		Consumption		Total Consumption
	N	P_2O_5	N	P_2O_5	N	P_2O_5	N + P_2O_5
Before 1960	Virtually none	Virtually none	Virtually none	Virtually none	-	-	-
1972	2,000	-	1,950	-	3,950	-	-
1973	2,400	725	1,800	428	4,200	1,153	5,353

[1]Source: "The Agricultural Situation in the People's Republic of China," Economic Research Service, ERS – Foreign 362, U.S. Department of Agriculture, May 1974.

Table 5. Computed estimates of sources of plant nutrients in the People's Republic of China, 1972, in millions of metric tons of nutrients[1]

	From Traditional Organic Fertilizer[2]	From Chemical Fertilizer	Total Plant Nutrient	Percentage of Plant Nutrient from Chemicals
High estimate	14.8	4.5	19.3	23
Low estimate	6.0	3.6	9.6	37

[1]China's Green Revolution, Benedict Stavis, Cornell University East Asia Papers #2, 1974.

[2]The plant nutrient content of the traditional organic fertilizer varies greatly according to Chinese scientists. The composition varies from: Nitrogen – 0.3 to 0.5%; P_2O_5 – 0.1 to 0.5%; K_2O – 0.2 to 1.0%. Therefore the estimates of plant nutrients of both the high and low level are at best a rough approximation of nutrients from this source. In Kirin we were told communes would like to apply 40-50 metric tons/hectare of organic fertilizer on corn but because of shortages generally can apply only 20-30 metric tons every second or third year. This explains the recent rapid increase in production and importation of chemical fertilizer and the recent decision to install ten (10) 1000/ton per day anhydrous ammonia plants and the corresponding urea conversion plant capacity.

Table 6. Increases in domestic fertilizer production in the
People's Republic of China, 1964-1972[1]
(millions of metric tons of fertilizer)[2]

Type of Factory	Year				
	1964	1965	1971	1972
1. Large nitrogen factories	2.0	–	2.0	2.1
2. Small nitrogen factories (mostly producing **ammonium** bicarbonate - 17%)	0.5	1.1	7.2	9.9
3. Small phosphate factories	1.3	–	2.8	3.0
4. Large phosphate factories	2.1	–	4.9	4.9
Total	5.9	8.9*	16.9	19.9

[1]Source: China's Green Revolution, Benedict Stavis, Cornell University East Asia Papers #2, 1974.

[2]An accurate estimate of the production of plant nutrients is difficult because most reports on fertilizer production refer to product without regard to percentage of plant nutrients involved.

*Total is known, but missing components are not. See footnotes to table 1.9, p. 41 of Stavis paper.

Table 7. Estimated harvested area, production, and yield of major grains in the United States and the People's Republic of China (average for period 1967-1971)[1]

Grain	Area Harvested (Millions of Hectares)		Production (Million Metric Tons)		Yield in Kilograms per Hectare		Chinese Yield as Percentage of U.S.A. Yield
	U.S.A.	China	U.S.A.	China	U.S.A.	China	
Rice	0.8	31.1	4.1[4]	92.0[4]	5000	3000	59
Wheat	20.6	24.4	41.2	23.6	2000	1000	49
Miscellaneous Grains	41.6	55.4	162.1	69.7	3900	1300	32
Tubers[3]	0.6	12.9	3.7	26.2	5900	2000	35
Total	63.6	123.8	211.1	211.5	-	-	-

[1]Source: Agriculture in the United States and the People's Republic of China, Foreign Agricultural Economic Report #94 (ERS) USDA, February 1974.

[2]Miscellaneous grain includes: millet, corn, grain sorghum, barley, buckwheat oats, beans, peas, rye, lentils, broad beans, field peas and other leguminous crops.

[3]Tubers: mainly sweet and Irish potatoes, converted to grain equivalents at one-fourth (1/4) of their actual farm weight.

[4]Unmilled paddy rice.

Table 8. The estimated harvested area, production and yield of major oilseeds in the United States and the People's Republic of China, average 1967-71

Oilseed	Area Harvested (Millions of Hectares)		Production (Million Metric Tons)		Yield in Kilograms per Hectare		Chinese Yield as Percentage of
	U.S.A.	China	U.S.A.	China	U.S.A.	China	U.S.A. Yield
Soybean	16.7	8.1	29.93	6.65	1,792	821	46%
Peanuts	4.5	2.0	1.23	2.41	2,460	1,205	49
Rapeseed	–	1.7	–	0.78	–	459	–
Sesame Seed	–	0.9	–	0.32	–	356	–
Sunflower	0.1	.05	0.10	0.07	1,000	1,400	140
Cotton Seed	4.2	4.6	3.67	3.28	874	713	82
Linseed	0.9	–	0.67	–	738	–	–

Table 9. Estimated livestock numbers in the People's Republic of China, 1949-1972[1]

Animal	1949	1972
	(1,000 Head)	
Large Animals[2]	59,775	95,042
Sheep and Goats	42,347	148,215
Hogs	57,752	259,884

[1]Source: "The Agricultural Situation in the People's Republic of China," Economic Research Service, ERS - Foreign 362, U.S. Department of Agriculture, May 1974.

[2]Includes cattle, water buffalo, horses, donkeys, mules and camels.

REPORTS ON SPECIFIC CROPS

In the eight years since the beginning of the Cultural Revolution, China has made no detailed statistical report on the land sown, yields, and production of the various crops. Over the past four years, however, many administrative units have issued reports of greater yields and production, expressed in terms of percentage increase over the previous year's crop. The lack of accurate, detailed, up-to-date production statistics makes it difficult to interpret change in either yield or production for particular crops or shifts in production between crops. Nevertheless, it is apparent that cereal and tuber crop production has increased substantially during the past seven years. In all probability the official claims of having gained self-sufficiency in 1971 were well founded. Reduced production in 1972 caused by droughts resulted in large imports. Although circumstantial evidence indicates that the 1974 harvest will be very good, and perhaps a new record, grain imports reportedly are continuing, almost certainly in replenishment of reserve stocks and compensation for reduction in chemical fertilizer imports. The latter resulted from soaring prices and shortages of fertilizers in international markets. Undoubtedly the need for fertilizer (independent of a vacillating world market) induced the decision to construct ten large anhydrous ammonia plants (1000 ton/day) and urea conversion units, assuring domestic food and fiber production.

Rice is the most important crop in the PRC. It is grown on a larger area than any other crop and the production is somewhere over three- to fourfold that of wheat, the second most important crop. In fact, the PRC, which is currently estimated to produce between 103 and 111 million metric tons of rice, is by far the largest rice producer in the world.

Wheat, the second most important food crop in terms of both area sown and production, is sown on 25 to 27 million hectares and estimates of the 1973 crop have varied between 28 million metric tons (USDA) to 42 million tons

(Stavis). Delegation discussions with scientists on the current trip give more credence to the USDA production estimates. The other grains (corn, sorghum, millet, barley, oats and rye) are lumped together in crop reporting. Collectively they produce much more tonnage than wheat. Among these, corn certainly occupies the greatest area and also provides the largest grain production. It is grown over a wide range of climatic conditions from Heilungkiang in the northeast (48° latitude) to the provinces of Kwangsi and Kwangtung and the island of Hainan in the south (18° latitude). It is apparent that maize is rapidly increasing in importance and appears to be outpacing sorghum production in many areas.

Rice

Rice, China's most important grain crop, is grown on about 34 million hectares of land which produce between 100-110 million tons of grain (paddy) annually. Commercial rice production and field research plots of rice were seen in each of the six provinces and municipalities visited, and were visible from the air or train in the other nine provinces through or over which we traveled.

We noted that in the north and northeast rice is relatively less important than wheat, maize and kaoliang. In the lower Yangtze river valley, it shares the fields with wheat and barley. In south and southeast China and the Szechwan Basin, rice is the primary grain.

In the regions north of the Yangtze river valley, japonica rices are grown; in the Yangtze river basin both japonica and indica rices are found. Irrigated areas of the Yangtze yield two crops of rice a year: the early crop is generally indica, the late crop, a day-length sensitive japonica type. South of the Yangtze Valley, indicas predominate.

From discussions with scientists, we gather that breeding for short statured indica varieties began about 1956 with the first semi-dwarf variety distributed in Kwangtung Province in 1960 (Kuang-chiang-ai is the variety name). The dwarf parent for this variety was said to be ai-Tze-Tzang from Kwangsi Province.

Varieties from the International Rice Research Institute (IRRI) appeared in China soon after they were released. IR-8, for example, released in 1966, was brought into China in 1967 and was tested in at least one province (Kwangtung) in 1968. Three years later it was said to have been tested in

Shensi Province. All other IRRI varieties have been evaluated in China including IR-5, IR-20, IR-22, IR-24, and IR-26. IR-26, which was released in November 1973, is already undergoing tests in China.

We were told that IRRI varieties all have one serious drawback for conditions in China; their growth duration is too long. Apparently the cooler temperatures lengthen to 145 to 170 days the time needed from sowing in the seed bed to harvest. China's emphasis on double or triple cropping has mandated the use of quick maturing rice varieties.

The crossing of IR-8 with local varieties in China was reported. Apparently the high yield and sturdy straw of this variety has made it a valuable parent in the breeding programs where indica types are grown. Indica lines currently being tested are likely to have IR-8 as one of the parents.

Rice disease and insect problems appear to be less severe in China than in the tropical rice growing areas. We were told that this was in part due to the fact that in the temperate north and northeast, cold winters and crop rotation tend to inhibit rice pests. Also, strict control is maintained on the movement of seeds of infected plants, seed treatment is in some cases made mandatory, and diseased plants are rogued from the fields. Furthermore, in some areas field monitoring and surveys are used to forecast pest build-up, making pesticide control easier. Pesticides are commonly used, but field monitoring makes it possible to use them only when needed. Disease and insect resistance is also a prime objective in rice breeding programs in all areas visited.

Chinese rice producers are concerned with pests similar to those that attack the crop elsewhere: green leafhopper (Nephotettix virescens), brown planthopper (Nilaparvata lugens), stem borer (Chilo suppressalis), leaf roller (Chaphalocrosis medinalis), and whorl maggot (Hydrellia philippina); diseases include blast (Pyricularia oryzae), bacterial leaf blight (Xanthomonas oryzae), and sheath blight (Corticium sasakii).

In general, fields were fairly weed-free, although some areas in the lower Yangtze river basin were less so than elsewhere. Weed control is accomplished by manual, mechanical, and chemical means. Chemicals used include MCPA, DCPA, 2,4-D, and Dalapon.

The most striking difference between cultural practices in China and those in tropical Asia is the number of seedlings planted per hill. Commonly five to ten seedlings were planted per hill, compared to one or two in many other countries. This practice was observed in every province we visited.

While this method may assure high main stem population, it does not utilize the high tillering capacity of some new varieties and is wasteful of seed.

Some evidence of mechanization of planting and harvesting was seen in the lower Yangtze river basin. Rice transplanters are said to be used on 20 percent of the rice land in the area around Shanghai; the primary objective is to speed up the planting of the crop and not to save labor. Similarly, a small mower-windrower was said to speed up harvest, making it possible to prepare the land more quickly for a subsequent crop.

Both organic and inorganic fertilizers are used for rice, though organic is more widely used than the inorganic. The organic compost materials are put on at planting, while at least some of the inorganics are used as top-dressers. Ammonium bicarbonate supplies much of the inorganic nitrogen, probably having been produced in one of the many small fertilizer plants found throughout agricultural China. Coal and water are the only raw materials needed for this fertilizer.

On farmers' fields there was evidence in most areas visited of a nitrogen shortage for rice. It appeared that some nitrogen had been applied, but leaf color often suggested that this was inadequate.

Research in the northeast confirms results obtained elsewhere that a concentration of nitrogen fertilizers near the root zone of the rice plant gave greater efficiency of utilization than broadcast applications.

The applied rice research and extension program which aims at integrating scientific and educational efforts into practical rice production programs is impressive. It appears to have greatly shortened the time span between discovery and practical application. Furthermore, the universal decentralization of research has encouraged initiative at all levels in the research system, with new varieties and innovative practices coming from production teams, production brigades, and communes, just as they come from academies, institutes, and colleges of agriculture. The experimentation base for rice has broadened greatly and is most impressive.

Decentralization appears to have brought with it, however, two features that merit concern. First, there appears to be much less coordination and communication among rice scientists than is desirable. On national and regional problems such as germ plasm collection and preservation, for example, a sense of coordination and direction is not easy to identify. Rice scientists with whom we talked seemed unaware of problems of germ plasm disappearance as new short statured rice varieties rapidly replace the landraces of the past. There seemed to be inadequate concern with this problem even on a regional basis.

Unfortunately, the publication of research results appears to have been suspended, at least temporarily. Failure to publish is creating a serious communication gap among rice scientists. They sometimes seem unaware of important work being done elsewhere and refer almost exclusively to work in their immediate geographic area; exceptions to this were the regional testing in the northeast and in the Yangtze river basin.

The second problem aggravated by decentralization is that research requiring long-term continuous inputs by a critical mass of highly qualified scientists tends to receive a low priority. At the provincial and lower administrative levels, highest priority must be given to those applied activities which will bring immediate practical benefits. Without in any way detracting from these remarkable applied programs, more fundamental research must be done. The need for centers or programs to accommodate this research is apparent. Discussions with some of the scientists suggest that they are aware of this problem, and they are convinced that already steps are being taken to strengthen and support the more fundamental research programs.

Germ plasm exchange and utilization pose China's greatest challenge and opportunity of the decades ahead. They have a real heritage of landraces of rice which they should collect, evaluate, and preserve. Through exchanges with other countries and the International Rice Research Institute, they could take advantage of germ plasm from all over the world. This diversity of genetic material will be needed to provide the future rice varieties China will need to feed its population.

The following are more specific comments on the rice areas and research units visited during our one-month stay in China.

During a visit to the Nan-yuan Peoples' Commune near Peking, we saw a good crop of rice which the manager suggested would yield between 6 and 7 metric tons per hectare. The excellent stand, good color, and freedom from weeds, and insect and disease damage indicate he may not be overestimating the probable yield. The rice, a variety called Tung Fang Hung, appeared to have at least moderate stem strength, was about 90 centimeters in height, but had only about 10 tillers per plant. It had been transplanted very thickly, a plant every 10 to 12 centimeters. Both organic and inorganic fertilizer had been applied, ammonium bicarbonate at the rate of 750 Kg/ha (135 Kg N/ha), and manure at the rate of 75 T/ha. If the manure contained only 0.3 percent N this rate would supply 225 Kg N/ha. If the indicated rates had been used, the plot was likely set up for demonstration purposes. BHC had been used as a pesticide.

Following a briefing at the Chinese Academy of Agricultural and Forestry Sciences (CAAFS), we were shown a rice experimental field. Nine varieties were being compared under two conditions: transplanted early in the spring; and transplanted following wheat. The latter plants were not as stiff strawed and were lodging. The varieties were the japonica type and about 100 to 110 centimeters tall. They were free of insect and disease damage and were well tillered. Mr. S. C. Lin, in charge of rice breeding there, estimated the plots would yield between seven and eight metric tons. They might well do so; these plots were the best we had yet seen on the trip.

Separate discussions with other Chinese scientists suggested that the germ plasm base for rice in the north, including the Peking area, is quite narrow, leaving considerable vulnerability in the event of a disease or insect outbreak. In general, however, the rice on commune fields looked good, although most of it was just coming into head, making it difficult to estimate probable yields. In most cases, plant and/or tiller population was high.

A report given by CAAFS representatives indicated that 80 percent of the rice grown in south China is short statured and stiff-strawed. If true, this area would have one of the highest rates of adoption for these varieties in the world.

Institute of Botany

Rice is one of the plants included in the anther culture work of this institute. Progenies of rice haploids (some which have undergone chromosome doubling) were quite uniform.

Institute of Genetics

Here we were shown further work on pollen culture and received publications (in English) describing the techniques. Millers medium supplemented with sucrose and 2,4-D or NAS increased callus formation. A second culturing in the absence of 2,4-D (with kinetin added) resulted in differentiation into green plants which were grown in pots to maturity. Half of these plants were diploids and capable of setting seed. F_2 and F_3 plants were reasonably uniform. This method, because it eliminates segregation, could shorten the breeding cycle, but as yet is not practical for large-scale breeding programs.

Kirin Academy of Agricultural Sciences

Returning from a visit to a commune, we saw the paddy variety test plots of KAAS. The objectives are to increase yields, reduce lodging, and enhance disease resistance. **The Bakanae disease was the primary concern. We saw an** early maturity test (120-125 days) of 14 varieties; the researcher suggested that the best of each variety would yield 6-7 T/ha. (This appeared to be a bit high.)

We learned that regional trials are run each year with rices grown according to maturity date. A team visits each plot during the year to make observations and recommendations for the coming year. Annual planning conferences are held.

NH_4NO_3 is said to be used in this area for paddy rice. A Chinese soil scientist agreed that much of the nitrate is likely lost by volatilization when the soil is flooded with water.

We were shown a replicated yield trial of 18 short lines and a nursery for screening resistance of 150 lines to Fusarium. They have about 500 varieties in their germ plasm collection; 200 are local and the remainder from Japan and Korea. They make 30 to 40 crosses per year.

They use DCPA as a contact spray to control barnyard grass. MCPA and propazine combined control Potamogeton and Calamus. NIP is also used as a herbicide.

During the discussion following the field visit we learned that one of ten large urea plants is to be built in Kirin Province. We were told that localized placement of fertilizer in peat balls gave higher rice yields than broadcast fertilizer applications. They have no machine to place fertilizer in a band near the transplanted rice. They said some ammonium bicarbonate is being produced in the province. Phosphorus deficiency was notable primarily in the western part of the province where soils with high pH are found. Potash is not now a problem but may be when cropping is intensified.

Rice has apparently been introduced into this province only in recent years. It is a relatively minor crop, however, due to limited water supply and short growing season. The scientists working on rice at KAAS appeared to be very knowledgeable and eager to cooperate if the opportunity arises.

Shensi Province

Rice is grown in only the southern portion of this province in the Han Shui river valley which is separated from the Wei Valley (in which Sian is located) by the Tsingling mountains. The rice in southern Shensi is of the indica type while the small amount grown in and around Sian is likely the japonica type. IR 8 which was said to have been introduced in 1971 has a growth duration (seeding to harvest) of 160 days in south Shensi - which is at least 10 days longer than desired. IR 8 is said to have yielded 9.0 metric tons per hectare - a very high yield.

Rice is grown as a part of three different cropping patterns. In one rice follows a fall planted green manure crop such as alfalfa, clover, or beans. The rice is seeded in April, is transplanted after one month, and is harvested in August. The other two cropping methods were not identified.

They indicated that insects and diseases were not major problems although blast was recognized as being at least a minor problem. Two short stature varieties - Nanking 11 and Short Pear - were identified and we were promised seeds of these varieties.

Nanking Agricultural Research Institute

This institute apparently plays a major role, not only as a provincial rice research organization, but as a regional coordinator of rice research in the Yangtze river basin. Scientists at this station help organize and monitor a series of coordinated yield trials and have the responsibility for collecting and maintaining germ plasm of the japonica varieties of rice.

The rice field plots were excellent. Both indica and japonica rices are grown here, but japonica varieties with photosensitivity do best in the fall season. None of the IRRI varieties have done well here because their growth duration exceeds the desirable limit (no more than 150 days). Apparently in this somewhat cooler climate these IRRI rices simply grow more slowly than under tropical conditions. They expect IR 20 to take 170 days from seeding to harvest.

We were shown a plot of IR 26 which was badly lodged under 210 Kg/ha of nitrogen. In contrast, the Chinese lines were standing and looked beautiful; the scientists estimate that they will yield about 7.5 T/ha. They had received the same fertilizer treatment as the IR 26.

The scientists appeared to be eager to cooperate and requested that IRRI provide them with seeds of a number of varieties. They want to receive materials with resistance to blast, bacterial leaf blight, and green leaf hoppers.

The plant protection scientists have developed an effective monitoring system for bacterial leaf blight, using a phage for detecting the organism. The

Chinese have been less successful than IRRI scientists in developing resistance to bacterial leaf blight, but some progress is being made.

Insecticides are in use as paddy sprays. Chinese scientists have obtained favorable results using the "mudball" technique of concentrating fertilizer near the plant roots. Similarly, root placement of insecticides (systemics) had been used successfully.

Shanghai Academy of Agricultural Sciences

This academy serves as the coordinator of a series of uniform trials of early-maturing japonica rice lines and varieties for 12 provinces and municipalities north of the Yangtze river. Annual meetings are held to discuss results and plan next year's trials.

IRRI varieties tried at this location were all unsuccessful (IR 8, IR 20, IR 22, IR 24, and IR 26) because their growing season (150 days and more) is too long to fit into the three-crop system common in this area.

Desired rices here are between 80 and 90 centimeters tall. The early ones, which are transplanted in late May, are generally indica while late rices, transplanted in early August, are japonicas.

They plant several seedlings per hill since they feel they cannot wait for tillering to occur. This practice may discourage tillering and certainly removes any disadvantage from varieties with little tillering ability.

There are few insect and disease problems in this area, blast being the most serious. They are using an organic phosphate (Kitazine) for blast control.

They are experimenting with the anther culture technique for rice breeding. We were shown the F_5 generation of a cross made using this technique. No publications are available on the method being used but discussions suggested that the methods were similar to those discussed at the Institute of Botany of the Academia Sinica in Peking.

Two machines to assist in rice mechanization were demonstrated. One, an experimental model, is for harvesting (cutting) rice. The rice is cut with a reciprocal blade similar to a hay mower and then moved to the side mechanically and left in a windrow. The second implement is for transplanting rice, and is operated by a three-man crew. One runs the 3 horsepower machine and two put the rice seedlings in place. Between one and two mou per hour can be transplanted with this equipment; by hand one worker transplants about 1/2 mou in eight hours. Thus the three people and the machine can transplant as rapidly as about 24 people planting by hand. We were told that this machine is used for about 20 percent of the late planted rice. The primary reason for the mechanization is to speed up the transplanting, not to save labor. The machine is said to cost 1000 Yuan ($500).

Kwangtung Province

Late rice as viewed from the train to Canton appeared to have been transplanted within the last two or three weeks. Except for some very recently transplanted areas, the fields looked green and showed no evidence of nitrogen deficiency. The fields were remarkably free of weeds and grass. Farmers were seen pushing small mechanical weeders and hand weeding for grass.

On a bus tour of a vegetable-rice area, we visited a commune field test plot having field trials of rice lines and plant density. Systematized rather than random arrangement of plots suggests these were more for demonstration purposes than research. The plots had some grass and other weeds, and the crop color was somewhat irregular, indicating a lack of adequate nitrogen. It may be that the vegetables were receiving more attention than the rice at this particular location.

Kwangtung Academy of Agricultural Sciences

The Grain Crops Research Institute of KwAAS, located a few kilometers from Canton, provides central research leadership in this province. Research is also done through brigade and team level field stations. This institute was formerly the Paddy Rice Research Institute of the academy.

We were given a brief history of the development of the short-statured, stiff-strawed indica rice varieties that are grown in this province. In 1956 a dwarf variety which had been found in a farmer's field in Kwangsi Province was obtained. The name of the farmer's variety was ai-Tze-Tzang. It was crossed in Kwangtung Province with a local variety of normal height in 1956. Four years later one of the selections from this cross, Kuang-ch'ang-ai was released. In 1961 the Chen-chu-ai variety was released and the variety called Canton Liberation #9 was released in 1964.

Apparently the spread of these varieties was rapid. By 1965, we were told, essentially all the early crop of rice was of the semi-dwarf type resulting from these crosses.

For the second, or late crop, however, adoption was somewhat slower since development of cold tolerance, resistance to disease, and day length sensitivity are needed. The late crop semi-dwarf variety Kwang-Er-ai was released

- 52 -

in 1963. It had yields similar to those of the early crop varieties and was winter hardy, reasonably resistant to diseases, and weakly photosensitive.

IR 8 from the International Rice Research Institute came into the province in 1967, one year after it was released in the Philippines. It was planted in Kwangtung Province in 1968. Unfortunately, its length of growing season in this area is too long (150-160 days) to fit the cropping patterns. Also, IR 8 was susceptible to bacterial leaf blight and was not cold-tolerant. Other IRRI rice varieties said to have been tested here include IR 20, IR 22, IR 24, and IR 26, and the line IR 661. None fit their growing season requirements.

We were told that IR 8 had been used in their breeding programs. They like its stiff straw and high-yielding ability. One would surmise that some of the lines currently being tested have IR 8 as one of the parents.

We noted the practice of planting seven or eight seedlings in one hill, a practice which we had observed in essentially every location visited in China. When quizzed as to why this was being done, they gave us no satisfactory answer. It was apparent that the practice was accepted and that no research had been done to ascertain its validity. With the coming of varieties with high tillering ability there is merit in checking out the need for such a dense planting. We suggested to the Chinese that they check on this practice.

The following dates of sowing, transplantation, and harvesting were given:

Operation	Early Crop	Late Crop
Sowing seed bed	Early March	June
Transplanting	Early April	Late July
Harvesting	July 15-30	Mid-November

Two major rice diseases are troublesome in this area: bacterial leaf blight (Xanthomonas oryzae) and sheath blight (Corticium sasakii). The latter was called oriental leaf sheath blight and identified as Pellicularia sasakii. Losses from bacterial leaf blight are minimized by strict quarantine on movement of rice seed from one infected field to another. Attempts are being made to breed for resistance to sheath blight.

We were surprised to learn that blast (Pyricularia oryzae) is not a serious disease in this area, although it is present.

Insect problems are said not to be too serious, although stem borer is a problem. Insecticides were being used (e.g., a compound called Samation) as sprays, but their use was dependent upon a rather rigid monitoring system to ascertain the presence of the borer. They had also tried Furadan, a systemic insecticide.

They indicated no knowledge of KwAAS's germ plasm bank's having regional responsibilities for the indica varieties for south China. This is at variance with what we had been told at Nanking in Kiangsu Province.

They claim to be making between 100 and 200 crosses annually at this station. This is the largest number reported to us for a single station. They showed us a field of F_3 lines and a high-yielding block which they estimated would give 7.5 T/ha. It looked good.

Wheat, Barley, Triticale

Late August and the month of September is not a good time to view or try to understand the research and production problems of the temperate climate small grains - wheat, barley, oats or rye - in the PRC. In the areas visited by the team, we did not see any of these crops growing.

The 1974 harvest of winter varieties was completed during the second half of May and the month of June. Spring sown varieties in the three northeastern provinces were harvested during the second half of July and the first week of August. There are small areas grown to spring habit varieties of wheat, oats, barley, and rye in higher elevations in some of the western provinces (harvested in August and September), but we did not visit any of these areas. Primarily, these plantings are important for local food needs.

The planting of winter wheat in the provinces of Hopei, Shantung, Kiangsu, Honan, Shensi, Hupei, and Anhwei does not begin until the last days of September and continues through the middle of October. Where winter barley is important it is planted about a month later than winter wheat. In some of these areas there has been a tendency in recent years to shift **to winter barley** rather than winter wheat, since barley ripens two weeks earlier than wheat, thus lending itself better to triple cropping (two crops of summer paddy and one of either winter cereal).

The spring wheat area is largely confined to the three northeastern provinces, Liaoning, Kirin, and Heilungkiang. Here the crop is planted as soon as the land can be worked into a satisfactory seed bed in early spring. Generally this is the last week of March through the first week of April, depending upon year and latitude.

Wheat Production

Currently wheat is grown to some extent in every province of China. The expansion into new areas in the south is the result of a government attempt to increase wheat production - as well as total food grain production - by including it in multiple cropping systems. Nevertheless, the great bulk of the wheat production today is derived, as it has been in the past, from the winter wheat areas of the Yangtze and Huang-ho (Yellow) river drainages, and from the spring wheat areas in the three northeast provinces.

Wheat is of far greater importance than barley, oats, and rye. In fact, the three latter crop species are of rather minor importance everywhere. Barley is, however, important as a winter crop near the coast in the Yangtze and Yellow river deltas where winter temperatures are more moderate and early ripening of the winter cereal crop is necessary if two crops of paddy are to be grown during the summer and fall.

Winter Wheat Production Zone

Winter habit varieties constitute from 80 to 85 percent of the total wheat production.

The northern limit of winter wheat cultivation varies from 40-41 1/2° N latitude in the north-northeast, depending upon nearness to the ocean (to ±35° N in the somewhat higher elevations on the Wei Ho, one of the main tributaries of the Yellow River, in the province of Shensi).

The northern limit of the winter wheat belt is determined by winter-kill. Winter killing increases and becomes more of a problem from east to west, and varieties with a considerably higher degree of cold resistance are necessary at Sian (Shensi Province) than at the same latitude on the coast of Kiangsu Province.

The southern limit of the winter wheat belt is less well defined than its northern limit. Moreover, the southern limit is now, or soon will be, modified by the effort to use triple cropping wherever feasible. This will necessitate the use of earlier maturing winter wheat varieties or, better yet, facultative types that can be isolated from spring x winter variety crosses if an imaginative program of breeding is organized.

The southern limit of the winter wheat belt roughly follows the Yangtze river flood plain, ±30° N latitude. Its southern economic limit, however, is somewhere between 28 1/2-30° N. South of this line presently available commercial winter wheat varieties either fail to head consistently, or if they do head, occupy the land too long. Under these conditions, the use of spring habit varieties is more feasible in a multiple cropping system. The southern limit of the winter wheat belt will almost certainly undergo considerable modification within the next decade as the campaign for more triple cropping gains momentum with the development of early maturing high-yielding crop varieties (including dwarf wheat varieties of a facultative growth habit) and greater availability of chemical fertilizer.

It is quite possible that some of the earlier maturing U.S. winter wheat varieties might be of value either directly as introductions or indirectly as parents in the winter wheat belt of the PRC. Materials from the U.S. Soft Red Winter, Hard Red Winter, and Soft White Winter Wheat Programs deserve testing. Consideration should be given to disease problems outlined elsewhere in this report in choosing materials for evaluation in the PRC.

Spring Wheat Production Zone

Most of the spring wheat (more than 80 percent) is grown in the three northeastern provinces (41-48° N latitude). In many ways this production zone is similar to that of North Dakota, Montana, Manitoba, and Saskatchewan. The principal difference is that the heaviest rainfall of the season occurs during late June and early July, when the wheat is ripening. This poses problems in control of certain diseases which are of little importance in the areas of the United States and Canada, and also greatly complicates harvesting and grain handling.

The introduction of the early maturing, high-yielding Mexican wheat varieties during the past three years has permitted expansion of fall-sown spring wheats into areas in the south where little or no wheat was formerly planted. Most of this anticipated expansion will be in the southern part of the conventional winter wheat belt, and in areas farther south, such as Hupei, Hunan, Kiangsi, Chekiang, Kwangsi, and Kwangtung. There are other local areas farther north where the Mexican type of spring wheat may be used successfully. Several hundred kilos of four Mexican spring wheat varieties were imported in 1972. During the 1973 crop season 5000 tons of several Mexican varieties of seed were imported and grown - apparently with considerable success in some areas. On the basis of the past two years of extensive testing (preceded by several years of small-scale tests using experimental samples of Mexican varieties obtained from Australia and Pakistan), the PRC has purchased the following quantities of seed for fall 1974 and spring 1975 sowing from Mexico:

Seed	Quantity
Potam	9,700 metric tons
Tanori	3,850 " "
Saric	1,100 " "
INIA	550 " "
Jori*	550 " "

We discussed the use of the Mexican varieties with a considerable number of scientists in different areas, but were unable to get any clear picture of where they will be grown commercially. Much of the seed is probably destined for the three northeastern provinces, and some for the extreme southern provinces where wheat is a new crop, commercially.

It is very possible that U.S. spring-wheat varieties such as Era and Fletcher may be valuable for use in the northeastern provinces.

<u>Wheat Research Programs Visited and Contacts Established</u>

<u>Spring Wheat Research in the Northeast</u>: The Kirin Academy of Agricultural and Forestry Sciences, Kung-chu-ling, Kirin.

- All wheats here are <u>Triticum vulgare</u>, of spring habit, planted March 25-April 15, and harvested July 10-20.

- Either white or red grain is acceptable.

- Good resistance to stem rust (<u>Puccinia graminis</u> tritici) is needed; it formerly was very destructive here. Now it is under control through use of resistant varieties.

- Leaf rust becomes heavy in some years. Tanori is resistant. There is no problem here with stripe rust.

- Glume blotch (<u>Septoria nodorum</u>) is a worsening problem. It causes serious losses in some years when rainfall is heavy and the atmosphere is humid and hot at the time of heading. Similar to Argentina in some years.

---

* <u>Triticum durum</u>.

-- Septoria leaf blotch (_Septoria tritici_) is present but never a serious problem.

- _Fusarium_ (_Gibberella_) may become a problem in the east (higher rainfall) in some years.

- _Helminthosporium_ leaf blight reduces yield seriously in some years.

Bunt (_Tilletia foetida_ and _T. caries_) formerly was a serious problem here but it is now controlled by seed treatment and resistant varieties.

Wheat yields prior to the 1950's were very unstable. Since 1968, with the release of the variety Fung-chiang #2 by the Kung-chu-ling station, this has been corrected. This variety has provided moderate to good protection against losses from the rusts, Septoria glume and leaf blotch (_S. nodorum_, _S. tritici_), and Helminthosporium.

- The Mexican varieties Tanori and Potam have outyielded Fung-chiang #2 by 30 percent in the past two years. Tanori seems to be the favored variety because of its combination of desirable characteristics. Both have adequate disease resistance, are short, are 5-7 days earlier in maturity, and resistant to lodging in the better rainfed or irrigated areas. Potam, however, tends to sprout when rainfall is heavy and temperature is high at harvest. About 100 Mexican wheat lines have been tested during the past two years. They were obtained via Australia. Many of them are high yielding and well adapted but need better resistance to glume blotch (_Septoria nodorum_) and _Helminthosporium_ leaf blight and to pre-harvest grain sprouting when temperatures are unusually high.

- This region is the only area in the PRC where the entire planting and harvesting operation is mechanized. Sowing rates of 170-200 Kg./hectare are common.

Organization of Kirin Breeding and Production program. Approximately 50 new F_1 crosses are made each year and two generations of breeding material are grown. The main season nursery is in Kirin and an off-season winter nursery is on Hainan Island in the south. Segregating populations are handled in a pedigree selection system through F_4 or F_5 when the best lines are bulked and included in yield tests.

The most promising lines from the preliminary yield tests in the Kirin, Heilungkiang, and Liaoning programs are evaluated in uniform yield trials in many locations throughout the ten provinces where spring wheats are grown.

An annual Spring Wheat Research Workers' Workshop is held each year where breeding methods, varietal recommendations, disease survey data, and agronomic practices are discussed and recommendations formulated.

Fertilization Practices: In the better rainfed or irrigated areas of the northeast, where organic manures are available, 20 to 30 tons/hectare are applied supplemented by 50 to 100 Kg. of N from ammonium nitrate. Where soils are deficient in phosphates, 50 to 80 Kg./Ha. of P_2O_5 are also included. Fertilizer practices are based on the past experience derived from many N-P-K performance tests in the region, rather than on chemical analysis.

Multiple Cropping and Intercropping with Spring Wheat. In the northeast spring wheat region (Liaoning, Kirin and Heilungkiang) the wheat harvest occurs too late to allow multiple cropping except on a very modest scale with early maturing crops such as summer vegetables, buckwheat, forages or green manure crops. Most of the land is fallowed after the wheat harvest.

During the past two years promising experimental results have been obtained by interplanting maize in spring wheat. In these experiments spring wheat was sown the last days of March and corn was sown in densely planted single rows widely spaced (several row spacings were studied) in late April. In these two years wheat yields reached 2 1/2 tons per hectare while the corn produced 6 tons per hectare. We saw the corn growing in these experiments and it was apparent that grain yields will be high even though plant height is considerably reduced. More research is needed to establish the commercial feasibility of this approach. In any case, usual means of mechanical wheat harvesting will not be possible with such a system.

In the southern part of Liaoning Province, where the growing season is considerably longer than in the two northernmost provinces, it may be possible to establish an economically feasible double cropping system using the early maturing Mexican spring wheat varieties and early maturing soybean varieties.

In the southern spring wheat areas, the early-maturing high-yielding Mexican varieties fit well into a triple cropping system of rice-rice-wheat. Aggressive wheat breeding programs will be needed to maintain adequate resistance to leaf and stem rust and to improve the level of resistance to scab (Gibberella), glume blotch (Septoria nodorum), and Helminthosporium leaf blight if this rotation is to be expanded and maintained successfully. The environment is often very favorable for severe epidemics of all of these wheat pathogens.

Wheat scientists in Kirin as well as Peking, Sian, Nanking and Canton had not heard of any of the international nurseries such as the USDA International Wheat Rust Nursery, the International Spring Wheat Yield Nursery, the International Winter Wheat Performance Nursery and several on durum, triticale and barley. All scientists indicated an interest in receiving copies of these various nursery reports, and scientists expressed interest in cooperative nurseries if government approval can be obtained.

Assuming that approval is forthcoming, the following nursery arrangements might be made at Chinese locations*:

USDA[1] International Spring Wheat Rust Nursery/Kung-chu-ling, Kirin and Canton, Kwangtung.

USDA[1] International Winter Wheat Rust Nursery/Sian, Shensi and Peking.

CIMMYT[1] International Spring Wheat Yield Nursery/Kung-chu-ling, Kirin and Canton, Kwangtung.

USDA[1] International Winter Wheat Performance Test/Sian, Shensi and Peking.

CIMMYT[1] International Spring Wheat Screening Nursery/Kung-chu-ling, Kirin and Canton, Kwangtung.

CIMMYT[1] International Durum Screening Nursery/Kung-chu-ling, Kirin.

Although no durum wheat is grown commercially in China, there have been several unsuccessful attempts since early 1900's to introduce its culture into the northeast. Before durum can be grown satisfactorily in these areas China must develop varieties with adequate levels of resistance to Septoria nodorum and resistance to sprouting under moist and high temperature conditions at the time of harvest. Several of the dwarf Mexican durum varieties, including Jori and Cocorit, are well adapted in the northeast but are extremely susceptible to both Septoria nodorum and sprouting.

---

[1] Coordinating agency

* At the research institute of the Academy of Agricultural and Forestry Sciences in each location.

It would seem desirable to have some of the durum varieties and lines from North Dakota submitted for evaluation.

- CIMMYT F_2 bulk unselected seed from spring x winter (T. vulgare) wheat varieties at Peking, Sian, Shensi; and Kung-chu-ling, Kirin.

- CIMMYT F_2 bulk unselected seed from spring x spring wheat crosses at Kung-chu-ling, Kirin; and Canton, Kwangtung. To be successful, segregates must be found that combine resistance to stem and leaf rust, Septoria nodorum, Helminthosporium leaf spot, Gibberella sp. (scab) and powdery mildew (in the south only). They should, if possible, provide segregates that are tolerant or resistant to grain shrivelling caused by hot dry winds when the grain is developing; i.e., resistant or tolerant to siroccos in North Africa or to arrebatamiento in Argentina.

- CIMMYT Triticale Screening Nursery at Peking.

- CIMMYT Barley Screening Nursery at Peking.

Winter Wheat Research Program at Peking: The National Academy of Agricultural and Forestry Sciences at Peking conducts research work on various aspects of winter wheat production at a number of its institutes. Most varietal improvement is done by the Institute of Genetics, though some undertaken by the Institute of Botany where pollen culture and chromosome doubling techniques are being studied. The quality of this rather basic research was good. There is also some work at the Institute of Mutation Genetics on mutation genetics of wheat; its quality was questionable.

Winter Wheat Research Program. The varieties for the winter wheat region must possess a desirable combination of the following characteristics:

- Adequate winter hardiness for the difficult wheat production areas.

- Resistance to major diseases, which include: stripe rust (P. striiformis) throughout belt; stem rust, especially in the southern part of the winter wheat belt and in the east near the coast; leaf rust (P. recondita) especially in the south; scab (Gibberella sp.) especially in the east and south; Helminthosporium leaf blight in the south and near the coast; and barley yellow dwarf virus.

- Resistance to lodging since much of the winter wheat is either irrigated or grown in zones of good rainfall and each year more fertilizer is being used. The first semi-dwarf Chinese winter wheats are being multiplied now, based on dwarfing genes obtained from crosses with Korean winter wheat varieties.

- Early ripening to permit wheat to be used in double cropping systems.

- Higher yield of grain.

- We heard no mention of industrial or nutritional quality being
  given any major emphasis.

From all indications the Institute of Genetics at Peking is carrying on a fairly large, well organized winter wheat breeding program, similar in both scope and organization to that described for Kirin Province. As in the spring wheat breeding programs, uniform province-wide and inter-provincial yield tests are conducted.

Agronomic and cultural practice experimentation is being conducted so as to fit wheat into more intensive multiple cropping systems. Many different approaches are being evaluated including intensive sequential multiple cropping in which winter wheat-rice, winter wheat-maize, or wheat-rice-vegetables are used to utilize the land effectively during the maximum number of growing days of each year. Intercropping in winter wheat is also being used in a semicommercial stage of development at present, to determine the feasibility of sowing either cotton or maize in winter wheat during early spring. Several thousand hectares of transplanted winter wheat have been grown the past two years. By this procedure it is hoped it will be possible to grow three crops during one year where only two are possible at present. In such a case winter wheat is being transplanted after the harvest of the second rice crop - more than a month later than it is possible to grow winter wheat successfully from direct sowing as has formerly been conventional. The limited results up to the present time appear promising.

Winter Wheat Research Program in the Northwestern Region at Wu-kung:
The Northwest College of Agriculture, where the majority of the reseach on winter wheat production problems is conducted, with collaborative work on some aspects of the problems being carried out at the Research Institute of the Academy of Agricultural and Forestry Sciences, is located a few kilometers from the College.

The research objectives are similar to those outlined above for the winter wheat program at Peking, except that certain modifications have been made because of differences in climate and disease problems. One major difference is the need for greater winter hardiness along the Wei valley (tributary of the Yellow) which lies at 500 meters elevation and where January minimum temperatures may drop to -17° C.

In this area a high level of stripe rust resistance must be incorporated into all new varieties. In some years leaf rust reaches epidemic proportions.

Leaf spots (a complex that probably includes Helminthosporium and Fusarium) severely attack many wheat varieties. Scab (Gibberella sp.), although not very important now, is likely to become much more of a problem as the intensity of corn usage increases in multiple cropping systems in which wheat is involved.

The first semi-dwarf winter wheat variety I-fung #3 has recently been released from the breeding program of the Agricultural College. The dwarfing gene was obtained from the Suewon (Korean) variety #86.

Winter Wheat Research at the Kiangsu Agricultural Research Institute: Department of Field Crops of the Academy of Agricultural and Forestry Sciences, Nanking, Kiangsu. Only a small wheat breeding program is carried on at this research institute in collaboration with the Agricultural College of Kiangsu. Kiangsu #1 is the most important commercial wheat variety in the province. The two most dangerous diseases in this area are stem rust (P. graminis) and scab (Gibberella). Scab and Septoria nodorum often become serious in low-lying areas where wheat is grown adjacent to paddy fields. The varieties grown here are of winter habit and mature in 200 to 210 days. Wheat, usually sown in late October or early November and maturing in late May or early June, is grown in a three crop rotation, i.e., wheat-paddy-paddy, or wheat-maize-paddy. The Mexican spring wheat varieties generally are not severely injured by frost in this area of the lower Yangtze and provide high yields of grain. When they are planted in bottom land adjacent to rice paddy fields, they are commonly hit hard by scab. The Chinese winter wheat varieties are also susceptible to scab. The best source of resistance to this disease is found in a Japanese variety that the Chinese call Nung-lin #50.

The wheat breeding program at this research institute is modest in both size and scope. The best contact at this research center concerning wheat research is Mr. Chou Chan-fei.

Winter Wheat Research: Shanghai Academy of Agricultural and Forestry Sciences. The work at this research center concentrates on breeding winter wheat varieties and spring barleys. Apparently the scope of both breeding programs is modest.

Wheat Research, Crop Research Institute: Kwangtung Academy of Agricultural Sciences, Canton. Until very recently very little wheat has been grown in this area, as it is subtropical. Now there is considerable interest in using wheat as the third crop in the triple cropping system with rice. In such a system, spring wheat would be planted in mid-November and harvested in early April. Mildew, leaf rust, stem rust and scab are all serious problems here. Some of the Mexican varieties such as Potam and Tanori are well adapted.

Madame Liu Shih-ching is in charge of the breeding program at this institute.

Wheat Disease Survey

A wheat disease survey is conducted each year in all provinces where wheat is important. Major emphasis is given to developing information and making collections of stem, leaf and stripe rusts. The collections and disease survey reports are sent to the three Regional Rust Research Laboratories where race identifications are made and the data are tabulated and summarized. These Rust Research Laboratories are located at: The Northwestern Agricultural College at Wu-kung (Sian), Shensi. Director of Plant Protection - Professor Li Chen-ch'i; The Agricultural University of North China at Chuo-hsien, Hopei; and The Agricultural College of Heilungkiang.

Annual meetings are held and summarized data are presented to wheat research workers in other disciplines. It was impossible to determine how effectively the rust race survey information is utilized to guide and improve the efficiency of the breeding and production programs. Many of these important questions remained unanswered because of the difficulties and slowness in communications via translators, and the short periods for discussions with the many scientists we met during the month we spent in the PRC.

The present system of research institutions, production brigades and peasant farmers is well suited to the rapid utilization of existing information developed in the basic sciences. There is some point, however, beyond which these practices will lose effectiveness; continuation of progress will require utilization of new developments from the supportive basic sciences of genetics, physiology, entomology, pathology, and others. We saw little evidence of the needed basic research in these fields.

Corn

The date of corn's introduction into China is unknown but possibly occurred sometime in the early 1500's. One avenue of entry was through the southwest as the first name applied to the crop was Jade Szechwan Panicum (sorghum). The common name for corn now translates as Jade Rice "rice" being a generic name for grain. In this early period it may have also been brought throught the ports of Macao, Fuchow, and Canton.

The first species were almost certainly the Caribbean tropical flints:
material now grown, particularly from Peking southward, still exhibits tropical
flint characteristics. In the intervening years, dent types were also introduced
and the introgression of dent germ plasm appears to have been most extensive in
the north. The dent corn influence is quite apparent in some of the hybrids due
to the use of a few U.S. inbred lines.

Planting

Corn planting follows a south-north progression similar to that in the
United States. Planting begins in Kwangtung, Kwangsi, and south Fukien about
mid-February to March 1. By April 1, planting is under way in Kiangsu, Hupei,
Honan, Kiangsi, and northern Fukien. By May 1-10, planting of corn, sorghum and
millet is underway in the northeast region. In all except the latter both an
early and a summer crop may be grown. Planting dates for the summer crop are
dependent on the crop succession pattern followed in a given region.

Planting rates are also variable depending upon fertility, water supply, and cropping pattern. When grown in solid stands or in combinations with
wheat, soybeans, sorghum or millet, planting rates of 40-60,000/ha are common.
With other types of interplanting, less dense planting may be employed.

Seedbed Preparation and Cultivation

Seedbeds are prepared by plowing, harrowing, or leveling. Horses and
bullocks provide the major power source though some small tractors are being
used as they become available. Planting is done by hand. Weed control is almost
exclusively done by hand hoeing. Herbicides are used to only a limited extent
because of problems posed by both interplanting and crop sequences.

Harvesting

Corn is harvested at high moisture contents, probably at physiologic
maturity. The plants are cut at ground level and piled in rows. The ears are
then husked and hauled to a drying floor. The stalks are hauled to some central
location and stored for winter feed.

After partial drying the ears are shelled and the grain is exposed to
further drying until it has reached a moisture level suitable for storage.

Breeding

Inbred line development is done by several agencies including the
Institute of Genetics of the Academy of Sciences, several branches of the
Academy of Agricultural and Forestry Sciences, provincial branch stations, and
some colleges of agriculture. The general pattern of operation is similar for
each organization.

Research staff conduct preliminary line development and evaluation. Combinations deemed to have merit are then evaluated by a cooperating production brigade of a local commune. The final decision as to which hybrid or hybrids to be grown (popularized) is made by the production brigade and the associated scientists from the research institute.

The parent lines are turned over to the production bridge; it then assumes responsibility for maintenance (purification), increase, and hybrid seed production. Where a commune has several production brigades, responsibilities for seed production may be shared.

Each commune attempts to satisfy its own seed requirements. No commune specializes in the production of seed for sale to other communes, but in emergencies seed may be shared among communes.

The first hybrids produced were top-crosses; current hybrids are either double-crosses or single-crosses with an increasing shift toward single-crosses. This shift appears to be conditioned as much by the greater number of isolations required for double-cross production as by any yield superiority of the single-crosses.

We were not able to see any inbred nurseries. From discussions it appears the methods employed were those in common use in the United States in the 1930's and 1940's. Inbreeding is practiced in local varieties. S_1 and subsequent generations are grown as ear-row progenies and selection is practiced within and among lines. Tests for combining ability are deferred until at least the S_5 generation. There appears to be no uniformity in test-cross evaluation procedures. Selection for disease or insect resistance is dependent upon the natural occurrence of these pests. An integrated approach involving the cooperative efforts of the breeder, entomologist, and pathologist has yet to be developed.

In the regions from Peking northward, commercial use is being made of a few U.S. lines. Some of those that are in commercial use or experimental plantings include M14, Oh43, C103, W75, W20, W24, and W135. With the exception of Oh43 and C103 these are lines of about 1940 vintage in the United States, where they have been replaced by newer and better lines. Seed of those inbred lines must have come to China by a circuitous route as the breeders having these lines knew little of their origin or special characteristics. Some of the newer U.S. lines should provide much superior combining ability as well as resistance to some of the important disease and insect pests.

In Kirin Province the major insect pest is the European corn borer (Ostrinia nubilalis) with only a single generation per year. Damage in 1974 was light. Only a limited effort has been devoted to search for sources of resistance.

The most important diseases are Helminthosporium turcicum (Northern leaf blight) locally called "big spot" disease, H. maydis (Southern leaf blight) and Spacelotheca reiliana (head smut). Other diseases observed that have potential for damage were Ustilago zeae, Gibberella zeae, Rhizoctonia sp. and Kabatiella zeae. Stalk rots are not particularly serious - probably due to the early harvesting.

We saw one planting in Kirin which had been artificially inoculated with H. turcicum. Some differences in degree of resistance were apparent, but none of the lines observed exhibited an acceptable level of resistance.

The lack of an effective arrangement for exchange of information and inbred lines is one of the major limitations in current maize programs. We were told meetings are held regularly, but there seemed to be little evidence of interchange of lines. There is at least one regional test of hybrids but the only items included are hybrids under production by the communes. A system for the production and evaluation of diallel single-crosses would be much more effective in furthering Chinese corn production.

Sorghum

All available information indicates that sorghum is an African crop, domesticated by Africans from the verticilliflorum race of wild sorghums at a unknown time. The most primitive domesticated race is called bicolor and is characterized by an open head, glumes one-half as long as the caryopsis to longer than the grain which is oval-oblong in shape and no more than slightly asymmetrical. The bicolor race closely resembles the wild races and is thought to be the progenitor of the more specialized guinea, caudatum, kafir, and durra races. The present distribution of the derived races is consistent with the idea of an early and wide distribution of bicolor from the northeast quadrant of Africa to West Africa, South Africa, and India. Remnants of bicolor are still found in India, Indochina and Indonesia. Evidence is tenuous, but it seems most likely that Indian bicolor moved northward into Yunnan and Szechwan and finally into north China where sorghum became one of the dominant cereals.

Chinese sorghums have developed certain unique characteristics. They retain the open bicolor head and seed shape, but the glumes are uniformly of the nervosum type, i.e., the lower glume is finely striated by numerous well-defined nerves and the upper glume is firm and shiny. The seed is tightly clasped by the glumes which in turn are firmly persistent. In these respects, they resemble broomcorn which, indeed, might be a kaoliang of some sort.

The Chinese words kaoliang (Gaoliang) or Ta Shu mean "tall millet" indicating sorghum was a later crop than the basic millets. For that matter, the north Indian word for sorghum, "Jowar" implies the derivative "reed barley," suggesting that sorghum arrived on the subcontinent after barley was widely cultivated in north India. The south Indian word "cholam" is unrelated. According to Professor P. T. Ho, University of Chicago, there was no word for sorghum in Chinese literature before the 4th century A.D., and the crop did not become sufficiently important to be taxed in kind until after the Mongol conquest. Nevertheless, reports appear in the literature of archaeological finds dating to Han times (200 B.C.-200 A.D.) or even to the Yang-shao Neolithic (3rd and 4th millenia B.C.). The reports have not been documented with adequate photographs, and considering the problems and difficulties of archaeobotanical identification and known misidentifications, they might not be reliable. The Neolithic dates in particular do not integrate with other information, but the presence of sorghum in China during Han times is quite plausible.

Nonetheless, sorghum appears to be a crop introduced into China, but introduced at a sufficiently early date that subraces have evolved that are characteristic of China and not found elsewhere except in nearby Japan and Korea. The Chinese sorghums or kaoliangs can be expected to have some unique properties. It is with considerable dismay, therefore, that we find the indigenous materials nearly completely replaced by hybrid grain sorghums based on the cytoplasmic male sterile system developed by Quinby and Stephens in Texas. Once again, unique and possibly important germ plasm has slipped away without any effort being made to conserve it in long-term collections.

Sorghum Breeding

The production of sorghum hybrids is similar to that for corn, but in general is simpler because less effort is put into the crop. This work may be done by provincial branches of the Academy of Agricultural and Forestry Sciences or by production brigades on the communes. At the Big Elm Tree Production Brigade, Kung-chu-ling, Kirin, we were shown the A + B lines and the R (restorer) line used. The A-B pair was relatively short, probably two dwarf genotypes and with a rather typical kafir-caudatum head and spikelet morphology. The glumes were typical of nervosum sorghums, however. The restorer line was tall with a more compact head than Chinese kaoliangs and had nervosum glumes. There was little evidence that more than a few male sterile lines had been set up and material in the field was rather uniform. Most of those examined in Kirin Province were tall, the heads much more compact than indigenous sorghums, the seeds decidedly asymmetrical like caudatum or caudatum-bicolor, but the glumes were universally nervosum.

A few fields or interplanted rows of short hybrids were also seen. The peasants in the communes like the tall types because they use the canes for fences, shelters, as a base for thatching, and other construction purposes. Kaoliangs are noted for tough, fibrous stalks and they have probably been selected for this quality for many centuries.

In Shensi, the hybrids were usually short in stature, but on the ir- rigated lands sorghum is rapidly giving way to corn which follows wheat in a double-cropping system. Wheat does not do as well after sorghum as after corn. The only breeding work seen in Shensi was done by the Academy of Agricultural and Forestry Sciences, Shensi Branch, and the crop had been harvested. The heads we saw were uniform in type (No. 6 heads, kafir-caudatum seed shapes, and nervosum glumes). Other organizations visited claimed sorghum was too un- important to work on. The area covered by sorghum in Shensi is very considerable; however, it appears to be policy to downgrade sorghum and promote corn at the present time.

Hybrids in both Kirin and Shensi usually had a brown seed coat. In Kirin sorghum is eaten more or less like rice. It is pearled to remove the seed- coats and boiled whole until tender. It makes a pleasant, palatable dish. The bran is fed to pigs. In Shensi, the cereal tradition is based on wheat. Sorghum is ground whole and baked into flat breads. The people complain about the bitter taste from the tannins in the brown seed coat. There is no reason for this; there are white-seeded kaoliangs available. They say the browns yield more, but this merely suggests they have not put much effort into a sorghum breeding program. The breeders have little conception of the range of materials that occur in sorghum - or most crops for that matter.

This was all the breeding work seen with sorghum. In the Nanking- Shanghai region, sorghum is grown in very small plots or in short rows along ditch banks or fences. These tiny plantings are all traditional kaoliang types that may well survive after the indigenous materials in the sorghum belt have been wiped out. They may be our ultimate hope for salvaging remnants of Chinese kaoliangs. Otherwise the indigenous materials appear doomed.

Sorghum Culture

The traditional sorghum belt of China is the northeast. Culture is very extensive today, although corn is increasing and at present the two crops are about equal in area. As the trend continues, especially under increasing irrigation, sorghum will decline still further. There is nothing unusual about the culture of sorghum, although a great deal of interplanting is now practiced in Kirin. Two to four rows of sorghum interplanted with two to six rows of soybeans are the most common configuration. Two rows of sorghum with four rows of soybeans is one of the most efficient, since both rows of sorghum are "outside" rows and receive maximum sunlight without appreciably intercepting light usable by the beans. North-south rows appear better at this latitude as there is too much shading on the north side of east-west rows.

From our observations, it would seem that every possible permutation and combination of interplanting is being tried out by the production teams. If there are real differences, a standard configuration will probably be adopted. A great deal of testing is done in this manner. Carefully controlled, statistically valid, replicated trials mean nothing if the masses do not like the practice.

Planting rates for sorghum appear to be less than optimum, especially under irrigation, but in Kirin they are interested in stalk development. In Shensi, there was some inconsistency in planting dates. Plots at the academy were planted in April and harvested in August. Most of the farmers in the communes planted after wheat harvest and harvested in September and October. It was obvious however, that sorghum in Shensi was not given the care and consideration of most of the other crops.

Finally, it should be noted that sorghum is the basis for Mao Tai, the 120-proof white lightning used for toasts at banquets. Without sorghum, Chinese diplomacy might suffer hard times.

Millet

Millet (Setaria italica) has been an important cereal in China for more than 6000 years. It is the first food grain grown by China's farmers. Its wild progenitor, Setaria viridis, is at home in Shensi Province and S. italica may well have evolved in this cradle of China's agriculture.

Today Setaria millet is grown in the semi-arid sections of the north and northeast regions. In Kirin Province, where it is the preferred food cereal, it occupies 15 percent of the cultivated land. The total acreage planted to millet in China is not known. In the irrigated Wei Ho Valley near Sian, corn has largely replaced millet. But when rainfed, millet will make grain where it is too dry for other cereals. Most of the land in China's semi-arid north and northeast cannot be irrigated. Here Setaria millet's future as a food and forage grass seems assured.

Use

Setaria millet, milled to remove the hulls, makes a delightful light golden porridge. It contains 12 percent protein and is considered superior to other cereals for winter food, especially for pregnant women. Flour from the glutinous varieties is used to make special sweet cakes for holiday celebrations such as the Spring Festival. These features plus its pleasant flavor make it easy to understand why Setaria millet is the preferred food cereal in Kirin Province where wheat, rice, corn and sorghum are also available.

Farmers and livestock specialists rate Setaria millet straw ahead of other cereal straws as a feed for livestock. Occasionally it is broadcast and cut before mature to make a high quality grass hay.

Culture

In Kirin Province Setaria millet is generally grown as single plants spaced 15 to 25 cm. apart in rows some 60 cm. apart. Here with adequate water and nutrients it may reach a height of 1.5 to 2.5 m. and have stalks 7 to 10 mm. in diameter. Its drooping head, up to 5 cm. in diameter, may reach a length of 40 cm. Although it is grown in pure stand, it is more frequently inter-planted with corn or other crops. Two to four rows of corn and six to eight rows of millet make a favorite intercropping pattern in the Kung-chu-ling area.

Near Peking and Sian, Setaria millet is broadcast to give single plants 15 to 30 cm. apart. Here mature plants rarely exceed one meter in height and heads are smaller.

Mature heads are usually harvested by hand from the standing crop and are spread out to dry before threshing. (On some roads, heads were spread out to be run over by traffic.) Six tons per hectare is considered a good grain yield. (Setaria millet outyields proso in the Kirin Province.)

As soon as the heads are harvested the plants are cut by hand, dried and stacked for hay. Harvested in this way and given good curing weather, the hay can be quite green and leafy.

Germ Plasm

Ms. Chin Lien-hua, Setaria breeder in the Kirin Academy of Agricultural Science at Kung-chu-ling, had small plots of 400 landraces of Setaria millet collected under field observation. Planted April 25, most landraces in the collection were maturing September 3, had large branched drooping heads and a height of 1.5 to 2.0 meters. At least one early maturing type had heads with little if any branching. Differences in head shape, plant height, plant color, and disease resistance were observed. Ms. Chin stated that dwarf types occurred but did not yield enough straw (forage).

Breeding

Setaria italica is a self-pollinated species, and Ms. Chin, the only millet breeder interviewed, has been breeding it since 1960. She has practiced intervarietal hybridization with F_3 to F_5 selections. Hybrids are made early in the morning by the hot water emasculation method. Heads ready to bloom are immersed for 10 minutes in thermos bottles of water at a temperature of 47 to 48°C. Dry heads are pollinated; this is continued on succeeding days if stigmas are exserted. Setaria millet produces an abundance of pollen.

Setaria F_1 hybrids may produce 20 to 40 percent more grain than their parents. Ms. Chin has never observed a male sterile millet plant. Hybridization of Setaria italica on female plants of other Setaria species was suggested as a possible means of creating a "cytoplasmic male sterile" breeding system. Irradiation of seed was also suggested.

Breeding objectives are increased yield of grain and straw, reduced lodging, and disease resistance. Rust, powdery mildew, and head smut are the worst diseases. Long bristles are sought as they reduce grain shattering. Both yellow and red glutinous varieties have been developed.

Ten new varieties were in testing at Kung-chu-ling and some of the 29 outlying stations in Kirin Province: No. 6, the most popular variety, a check; No. 23, greatest disease resistance; and No. 29, highest grain and straw yields but more lodging. Dwarfs tested produced less grain and straw. Dr. Burton described his work with dwarf pearl millet, suggesting that shortening the straw would reduce lodging and increase leafiness and enough straw quality to offset the loss in dry matter yield.

Full season Setaria millets require 135 days to mature at Kung-chu-ling (43°N latitude). In the extreme south (20° latitude), they mature in 90 days.

Dr. Burton indicated our desire to exchange germ plasm, listing especially preferred strains. Several were given to us on leaving the academy.

Panicum miliaceum

Proso millet, _Panicum miliaceum_, appears to be of little economic importance in China. Ms. Chin reported that it yields less than Setaria millet and
livestock will not eat the straw. She said some people grow a little to make
glutinous flour but that her glutinous varieties of Setaria millet are equal in
flour quality and superior otherwise. She was not doing any work with this species. We saw very little proso millet in the parts of China we visited.

Pennisetum americanum

Pearl millet, _Pennisetum americanum_, is also of little importance in
China. Ms. Chin had one strain in her millet collection with well filled heads
about 9 feet tall. She stated that it does not yield as well as Setaria millet,
but said she would be glad to grow several pearl millets that Dr. Burton will
send her. She may be able to get some collections of pearl millet from those
areas where it is grown in a very limited way. Dr. Burton did not see any pearl
millet elsewhere and no other person he talked with had heard of it.

Soybeans and Other Grain Legumes

The soybean is very widely grown in China, but only in the three northeastern provinces is it considered to be a major commercial crop. Here large
amounts of soybeans are sold to the state and made into traditional soy food
products (soy sauce, various bean curd products, etc.) for sale all over China.
These represent an important high-protein part of the diet, and are in such demand in the Shanghai area that soy products are the main food items rationed in
the markets there.

Soybeans may be seen growing in small fields, gardens and waste places
in most of eastern China. In Peking and Sian areas, the ripe seeds are used in
food products, usually by the grower himself or at least within the commune. In
Nanking, Shanghai, and Canton, in addition to the above use, the soybean is consumed as a green vegetable, and large quantities are sold in the market as green
pods. In some areas the soybean is used for livestock feed, the whole plant or
just the seeds, and it is often grown for green manure.

In the northeast, because of the shortness of the season, soybeans are
not double-cropped but are planted in the spring in late April or May and harvested in late September and October. They are seeded five cm. deep on top of
ridges that are formed about 20 cm. high and about 60 cm. apart. The ridge is
to keep excess water from the seedling and to keep the soil warmer in the spring.

- 73 -

A good plant population is considered to be about 170,000 plants per hectare, considerably lower than in the United States. Planting is by hand or planter with hand-thinning. Fertilizer is applied in the form of 30 to 40 tons of manure per hectare. Irrigation is rarely used with soybeans. Hand-weeding is the common practice but on highly mechanized farms some herbicide may be used. Soybeans are usually harvested with a hand sickle, and after a little drying are threshed with an electrically powered cylinder thresher. In the Kung-chu-ling and Chang-chun area in central Kirin Province, virtually all soybeans are intercropped with corn; six- or eight-row strips of soybeans between much wider strips of corn.

In the rest of eastern China soybean growing methods are as varied as the climate, topography, and cropping systems into which they must fit. Soybeans, more than any other crop, are used to fill in what otherwise would be wasted land. They are seeded along ditch banks, stream banks, terrace walls, roadsides, railroad sides, paddy dikes, field margins and any other odd-shaped piece of ground too small or too steep for a commercial crop. Weeds are clipped down by hand or by grazing livestock. Stands and growth are often poor, but some nutritious food or feed is produced from land which would not otherwise be used. In addition, there are small fields (1 to 4 hectares) of soybeans in most areas and they are often intercropped with corn or sorghum in the north or with fruit trees, bananas, mulberries, etc. in the south. Everywhere it is seen in the small gardens around farm homes - often the predominant species in the garden. Soybeans may be double-cropped with winter grains in the north. In the south two soybean crops are possible, one planted in March and harvested in mid-summer, and the other planted in June or July and harvested in the fall.

The major soybean breeding center visited was at the Kirin Academy of Agricultural Sciences at Kung-chu-ling. Two other breeding centers were also visited; the Genetics Institute in Peking, and the Northwest College of Agriculture west of Sian in Shensi Province. There are also breeding centers in the two other northeastern provinces, Heilungkiang and Liaoning, but apparently only a few small breeding or testing programs elsewhere in China.

Soybean breeding has evolved since 1949 when local varieties were gathered, evaluated, and the superior ones popularized to the present stage of developing improved varieties through intervarietal hybridization followed by pedigree selection. Approximately 80 percent of the soybean area in Kirin Province is sown with these improved varieties developed at the academies or agricultural colleges. In other parts of China where soybeans are less of a commercial crop and where less breeding is done by the researchers, local varieties presumably are much more predominant.

Kirin local varieties ranged from 110 to 140 days in growing period and from 10 to 20 grams per 100 in seed weight. The predominant plant type is indeterminate, with semideterminate next, and determinate the least frequent type. Varieties currently grown in Kirin predominantly have white flowers and gray pubescence. About one-third have narrow leaves ("to get better light penetration" but they did not feel that seed yield was necessarily higher).

Intervarietal crossing has involved types from Kirin and other provinces in China. Pollination is done on a rather unusual time schedule with emasculation beginning at 5 a.m. and pollination done from 8 to 10 a.m. After 10 a.m., the pollen is gone. A winter nursery is grown on Hainan Island where the F_1 and alternate generations (F_3, F_5, etc.) are produced to hasten breeding.

There was about one hectare of F_2 populations being grown at Kung-chu-ling involving some 50 different combinations. The F_5 to F_8 generations are grown in progeny rows at Kung-chu-ling (1300 of them in 1974) and visually selected lines are performance-tested in subsequent years.

Final strain testing is done in a regional test grown here and at over 30 locations in the province. In 1974 the test consisted of 25 varieties including experimental strains developed at Kung-chu-ling and five district research institutes. The test at Kung-chu-ling is at two population rates in plots five rows wide spaced 60 cm. apart and 10 m. long, in four replications with one replication at a higher fertility level. Expected yield level is 2000 to 3500 kg/ha. There were 11 U.S. varieties of maturity groups I to IV being tested here for the first time: Amsoy 71, Beeson, Calland, Clark 63, Corsoy, Harosoy 63, Kanrich, SRF 307, Wayne, Wilkin, and Williams. Growth was very good with a height of nearly four feet for the taller varieties. There was no lodging, partly because of the low population of 170,000 plants per hectare. These varieties looked similar to U.S. counterparts, except that Williams had rugose leaves, was stunted, and poorly podded.

In addition to intervarietal hybridization a mutation-breeding program is under way using x-rays or gamma rays in a search for earlier maturity. Some progeny rows in the R_5 generation appeared to be earlier than the check variety.

Objectives of the breeding program are: high yield, high oil content, resistance to pests, adaptation to intercropping, and strong stem. Eight improved varieties have been developed here and popularized in Kirin Province and nearby areas.

The major pest problem is the soybean pod borer. If fields are not treated with insecticide the borer will infest 10 to 20 percent of the pods. Moderate resistance has been found in the variety T'ieh-chia-shi-li-huang, local to Kirin. Soybean mosaic is not considered important in the breeding program although plants with mosaic-like symptoms and poor podding occurred frequently in the field at Kung-chu-ling, and this appeared to appreciably affect yield. Aphids are a major insect problem but resistant varieties have not been found.

At Kung-chu-ling, we saw the only cultural research on soybeans, an experiment in intercropping with corn. Six-row strips of soybeans alternated with corn were being tested against systems in which each crop was grown alone.

The soybean breeding program at the Genetics Institute, Peking, was begun in 1968, and has emphasized disease resistance. The varieties Ching-hua 3 and Ku-huan 2 were selected here, tested on county experiment farms, and are beginning to be grown by farmers. They are resistant to purple stain (Cercospora kikuchii). The main disease problems now are three viruses, soybean mosaic (the most serious one), soybean stunt, and bud blight (caused by tobacco ring spot virus), and the leaf disease target spot (Corynespora cassicola). A display plant specimen infected with soybean mosaic was very stunted and almost podless, much more severe symptoms than those associated with the disease in the United States. Resistance has been identified by observing plants in the field for symptoms and checked by inoculation tests and the electron microscope. Two varieties were found to be moderately resistant, Tung-hsien and Kirin No. 3, and two to be immune, Chi-hua 1 (a local variety from Shantung Province) and Shu-chou 424.

Now that scientists at the Institute of Genetics have the needed disease resistance the next step will be to breed for yield, and lodging resistance. They are growing both determinate varieties, which are lodging resistant and grow well in the lowland, and indeterminate varieties which do relatively well in poor soil areas and are grown in the upland. In the upland area north of Peking, soybeans of an indeterminate type and medium to small seeded are often interplanted with corn or sorghum. However, growth and yield prospects did not appear good where soybeans were in close competition with the taller crop. Occasionally there were small areas planted solely to soybeans; in these the yield prospect appeared much better.

Soybean breeding at the Northwest College of Agriculture west of Sian

was begun more than ten years ago. Current research is carried out in cooperation
with the nearby Academy of Agriculture - the source of germ plasm. The first objective is to develop early maturing varieties which can be planted after wheat
harvest in mid-June and ripen in early to mid-October. A second objective is
disease resistance, mainly to an unidentified virus which causes stunting of the
plant and wrinkling of the leaves, perhaps the one identified as soybean mosaic
at the Institute of Genetics. The only other disease considered is bacterial
leaf blight (Pseudomonas glycinea). Pod borer (Encorma glycinivorella) is the
most important insect pest but no adequate resistance has been found.

In 1970 this institution developed a new variety, Niu Mao Huang, by
pedigree selection from the cross Yi-wo-feng (a local variety) x Pin-ting-huan
(a Shantung variety). It is determinate, grows about 70 to 80 cm. tall when
planted after wheat, matures early in 105 to 110 days, is shattering resistant,
and virus resistant. This variety can be successfully double-cropped after
winter wheat; this has become a successful practice in central Shensi. Local
soybean varieties in Shensi Province are mostly determinate and are used for
bean curd or animal fodder (cattle and pig). There are also a number of black-
seeded indeterminate varieties which are grown for fodder; the threshed grain
may be fed to animals.

There does not appear to be much soybean research in other parts of
China. Yet, extensive soybean growth throughout the country and the importance
of the crop as a source of protein, especially for rural people, should justify
more research. In the Nanking-Shanghai area, all soybeans observed were determinate and very large-seeded. Fresh green soybeans in the pod were a common item
in the markets. In the south in Kwangtung Province many of the soybeans were determinate and of moderate seed size but some fields were planted with long-
duration, indeterminate varieties.

Peanut culture and research is apparently concentrated in Shantung
Province although small- and medium-sized fields are common in Kwangtung Province.
Nothing was learned about peanut research except that it is mainly conducted at
the Shantung Academy of Agriculture in Shantung Province.

Other grain legumes such as cowpea and mungbean were often seen interplanted with soybeans or instead of soybeans with other crops, in the wayside
areas, and in gardens. Mungbeans are used for sprouts and the cowpeas are eaten
as a green-pod vegetable or as boiled seeds, depending on the variety. No research information was found on these crops although the long-podded cowpea and

occasionally the common bean and hyacinth bean (Dolichos lablab) are grown on a large scale by the vegetable-producing communes.

One source of soybean germ plasm that is apparently not being used in China is the wild soybean (Glycine soja). It is found in many areas but was observed to be more abundant in the northeast. Perhaps it has been eradicated from the other agricultural areas of China since it was not seen in the Peking, Sian, and Canton areas and was found in the Nanking-Shanghai area, only in nonagricultural areas such as a forest park and wasteland around a factory and an airport.

At the Institute of Botany of the Academy of Sciences at Peking, the herbarium has a collection of about 100 sheets of Glycine soja from the following provinces or autonomous regions: Heilungkiang, Kirin, Liaoning, Neimengku (Inner Mongolia), Hopei, Shantung, Honan, Shansi, Shensi, Kansu, Kiangsu, Anhwei, Chekiang, Szechwan, Kiangsi, Hunan, and Kwangtung.

At the herbarium of the Sun Yat-sen University in Canton, there were a few old sheets of G. soja, G. pinnata (now considered to be Ophrestia pinnata), and one of G. hainanensis. This last, collected in 1933 on Hainan Island was a most interesting specimen. It was listed as an annual, and judging from the seeds and pods it is very close to G. soja although botanists call it Teyleria koordersii. The pods occur in rather large clusters and have up to seven seeds per pod which would make it a very interesting relative of G. soja. Efforts should be made to collect this species for further study and as potential germ plasm for soybean breeding.

Vegetables

Reports of a number of agricultural visitors to China since 1972 have been notably lacking in references to vegetable production. Actually, one cannot enter a city in China by any means of transportation in daylight without seeing large areas devoted to vegetable growing. Our hosts of the Agricultural Association assured us at the outset that vegetables are important in the Chinese diet, more so than meat. They said that most Chinese eat a little meat each day but eat a large quantity of vegetables. They estimated average consumption at one-half kilo per person per day. Observation of prices in the markets and the baskets of the shoppers confirms this.

Vegetables are grown very close to or even within the cities. Many are still delivered to market by handcart. There is almost no processing, and we were told of no specialized production region that would supply a particular vegetable to several cities. Therefore the population of a city is dependent on the vegetables from the fields nearby. There is a limited amount of storage to prolong the marketing season, especially in the north.

September proved to be a surprisingly poor month to see vegetables. Even though our travels covered latitudes corresponding to the range from northern New York to southern Florida, we consistently heard that summer crops were finished and many fall crops just planted. Even though there are over 100 kinds of vegetables grown in China, we saw only a small portion of these. Chinese cabbage in its various forms was by far the predominant vegetable in the fields.

There seems to be a deprecatory attitude towards vegetables in China which may be worth mentioning at the outset and may explain the lack of reference to them by other agricultural visitors. At the Kirin Academy of Agricultural Science we were told that vegetables were not very important in that area even though we could see solid areas of vegetables around the cities. Throughout China we have been told that horticulturists do not work on sweet potatoes because they are a food rather than a vegetable. In Canton, we heard that "white potatoes are a food in the north, but here we consider them a vegetable." In the Shanghai area we saw many small plantings of soybeans, but horticulturists said they did not work with them because here they are used only as a green vegetable. Horticulturists seem to have little awareness of the nutritional contribution of vegetables and repeated inquiries gave no good leads to any persons with such information. However, on the basis of studies in other developing countries, it seems likely that vegetables are the major source of vitamins A and C and probably some of the minerals.

Before describing vegetable production in various areas, it may be well to explain terminology related to Chinese cabbage since this caused confusion during much of the trip. What will be called Chinese cabbage is a closely related group of vegetables having different names in Chinese without adequate English equivalents. In the north, the name refers to the large heading type, grown also in the United States and called in China Brassica pekinensis, big-headed cabbage. As one moves south, an increasing amount is the nonheading group, Brassica chinensis, with its common name translated as "small-headed cabbage." In the Canton area, a third group is widely grown and distinct in that it flowers very early and is eaten as a 30-day old plant; all of it - leaves, elongated stem, and buds or flowers. We were told that this is B. parachinensis, with no English equivalent. Because of their interfertility,

<u>B</u>. <u>pekinensis</u>, <u>B</u>. <u>chinensis</u>, and <u>B</u>. <u>parachinensis</u> are considered by some in the U.S. to fall within the species <u>Brassica</u> <u>campestris</u>. In most of the following discussion the three will be grouped as Chinese cabbage.

Observations on Commercial Vegetable Production

The Nan-yuan People's Commune on the outskirts of Peking has over 41,000 people, 2100 hectares of land, 16 production brigades, and 135 production teams. Its main products are vegetables (90 million kg in 1973) and grains (7.5 million kg in 1973). Thirty-four thousand pigs are fed with vegetable waste supplemented by some grain. About 1100 hectares of land are used for vegetables and 1000 for wheat and rice plus some maize and sorghum.

The statement was made that over 80 "varieties" of vegetables are grown by this commune. Summer crops had largely been harvested and fall and winter crops were in early stages of development. An unusual eggplant with flattened fruit was about the only warm season crop still producing. This was one of five eggplant varieties grown, this being used for the late crop. We were told that tomato, pepper, and cucumber are grown, but none were seen. At dinner later we were told that tomatoes and watermelons had been harvested. With at least a month until frost it would seem that disease may be shortening the productive life of these crops or else that the land is needed more urgently to get fall crops started.

Yardlong beans were ripening on poles at the edges of fields and ditches, apparently to be used for seed. Much bean rust (Uromyces phaseoli) and mosaic were evident on these plantings.

Chinese cabbage was the most prevalent vegetable being planted in dooryards as well as in fields. In some fields the heading type was growing on ridges while non-heading type was growing thickly in furrows. Apparently the latter was being thinned from time to time and the plants removed were eaten or marketed. Alternate beds, perhaps 20 feet wide, were planted with leeks. The leeks would be harvested at monthly intervals beginning in February. An alternative to beds is to provide space for windbreaks which are constructed by piling up the soil in the beds from which Chinese cabbage has been harvested. A number of beds were observed in which spinach was just emerging.

The five vegetable specialists with whom we had a discussion on September 2 were asked about the relative importance of various vegetables in China in terms of land used. The group listed the following as the top four in order of rank: tomato, cucumber, Chinese cabbage, and cabbage.

Vegetables in Kirin Province

In the vicinity of Kung-chu-ling, there were many fields of Chinese cabbage, cabbage, and onion, along with smaller plantings of a great many other

vegetables. On the road from Kung-chu-ling to Changchun there were fields of
the same three vegetables plus eggplant, turnip, potatoes, common beans, and
some sweet potatoes. Chinese cabbage was by far the most abundant vegetable,
probably followed by eggplant. The onions all appeared to be Chinese, or non-
bulbing, onions (Allium fistulosum). Both tomatoes and cucumbers were conspicu-
ously absent.

Dr. Munger had a chance opportunity to walk briefly in two vegetable
plantings of perhaps one-quarter to one-half hectare each, apparently used by
institutions. One was next to the fields of the Big Elm Tree Production Brigade
and the other associated with the Fruit Research Institute. In both, the vegetables were rather poor in growth and gave evidence of poor seed. Young turnips
showed extreme variation in leaf type, bell peppers were misshapen and corrugated,
and cabbage had numerous plants with no heads mixed among others with heads approaching maturity and some bursting. While these faults may not have been too
serious in these small plantings, they are indicative of an inferiority of seed
which could include characteristics less readily observed but important such as
yield ability and resistance to pests.

Areas we saw that were devoted to vegetables would justify much more
research.

A Private Vegetable Garden at the Big Elm Tree Brigade, Kung-chu-ling, Kirin Province

No mention was made of vegetable production at this commune, but while
visiting the fields we did have an opportunity to see a vegetable planting that
we were told was a private family garden. The crops were potatoes, turnips, mung
beans, and Chinese cabbage, the last making up about 80 percent of the planting.
The Chinese cabbage and turnips were relatively young and it is likely that they
had been preceded by other vegetables.

The potatoes had extremely poor vine growth and some virus disease was
evident. Their productivity must have been quite low. The turnips showed extreme variability in leaf type, and there was much variation in the Chinese cabbage. It would seem from these observations that the quality of seed was not
good.

We thought that the amount of Chinese cabbage was too great for one
family but were told that it can be stored in pits until March or April, and
that in cold weather it is also preserved in large containers of water where
it develops a sour flavor that is appreciated.

Vegetable Production in Shensi Province near Sian

On September 9 we visited the Five Star Brigade of the Red Flag Commune in a suburb of Sian. About 53 hectares of vegetables constitute one of their main sources of income. They are growing about 200 varieties of 60 different vegetables. The brigade has nine production teams and an experiment station where two vegetable specialists from the Shensi Academy of Agricultural Sciences are currently stationed. This brigade produces most of its own vegetable seed with enough surplus to supply some other brigades. They credit the breeding of more suitable varieties in their own commune with substantial yield increases. They have also increased intercropping so that three or four crops are grown on the same land each year. This has also increased yields per hectare. Both cucumber and winter melons are grown on trellises or frames to increase efficiency of land use. Rain and approaching darkness prevented us from seeing the production fields.

Mr. Chao Shih-yah, a leading member of the Vegetable Research Group of the Shensi Academy of Agricultural and Forestry Sciences and one of the specialists at the Five Star Brigade, gave the following lists of vegetables in order of importance in Shensi Province:

Spring: tomato, cucumber, common bean, cabbage, cauliflower

Autumn: Chinese cabbage, cabbage, carrot, eggplant, potato

The following vegetables, not in order of frequency, were seen in the fields during drives south of Sian to the Fruit Research Institute in Mei County and west to the Northwest College of Agriculture at Wu-kung: Chinese cabbage, cabbage, onions (non-bulbing), leeks, carrots, sweet potatoes, peppers (small, red hot peppers being semi-dried as well as many still in the fields).

Vegetables in Kiangsu Province

We had no opportunity to see vegetable fields in this province. Mr. Wu Kwan-yuan of the Agricultural Research Institute provided the following lists of important vegetables, probably not in order of importance:

Summer: tomato, eggplant, pepper (hot and sweet), pole bean, cucumber, wax gourd

Winter: spinach, Chinese cabbage, cabbage, leek, turnip, stored radish, stored onion*

---

* Maleic hydrazide treatment to prevent sprouting was said to be common for stored onion.

Vegetables at the Rainbow Bridge People's Commune, Shanghai

This commune cultivates 3,120 hectares and has 26,647 people in 6687 households. The land devoted to vegetables produced over four crops per year; in 1973 the total yield was 84 tons/ha. Seventy-three thousand tons of vegetables were supplied to the urban markets, providing vegetables for 500,000 people or one-twelfth the population of Shanghai. (This amounts to 400 g/person/day; reasonably close to Mr. Huang's figure of 500 g.)

It was stated that 168 different vegetables are grown at this commune but the number we saw or heard mentioned was relatively small. Of course we saw only a small portion of the commune and had little time to get to the fields. Crops we saw or heard about included: pepper, eggplant, tomato, Chinese cabbage (non-heading), ginger, mushroom, turnip (intercropped with Chinese cabbage), water convovulus, stem lettuce, celery (Chinese cabbage to be interplanted later), carrot, radish, potato*, cauliflower, garlic, scallion, cabbage, and sweet potato.

We saw celery being transplanted for harvest at the end of October. Chinese cabbage was to be interplanted later. The celery was planted very close, rows were estimated at 30 cm. apart and plants 6 to 7 cm. apart. Judging from the spacing, the time until harvest, and the size of celery seen in the market, the celery is harvested before its period of most rapid growth is reached. It would seem in order to conduct experiments to determine the spacing and length of growing period that would produce the most celery per unit of land per day the land is used.

Much blight was present on the celery leaves at transplanting. It is difficult to believe that this is not hurting production.

We were in two fields of cabbage. One had been intercropped in a field of yardlong beans, was planted about 10 cm x 10 cm, and was suffering from both downy mildew and worm damage. In addition there was evidence of great genetic variability in plant size and maturity. As a result, we were told this crop would be fed to the pigs. A better field but one still showing much variability was planted 45 x 45 cm. In cabbage, close spacing delays maturity and amplifies the effects of genetic variability. It would be desirable to have spacing studies conducted with seedstocks being used as it appears that they may have gone too far in the direction of close spacing.

---

* Two-crop systems used for producing seed locally.

A field of trellis tomatoes with the second cluster open showed much evidence of virus. They expect to get 10 to 12 ripe fruits per plant before frost in mid-November. They said that 2,4-D is being applied to improve fruit set. This was the first tomato planting of any size that we saw.

At the Shanghai Academy of Agricultural Science we were given the following list as the order of importance of vegetables in this area:

Spring: non-heading Chinese cabbage, tomato, winter melon, cabbage, eggplant, snap beans, potato

Fall: non-heading Chinese cabbage, heading Chinese cabbage, spinach, celery, turnip

Vegetables at the Hsin-ch'iao Commune, Canton

This commune located almost within sight of Sun Yat-sen University has 58,000 people, 3467 hectares, 19 brigades, with 9 growing 700 hectares of vegetables, and 220 production teams. They grow 60-70 different vegetables. About 20 were in the fields at the time of our visit. Two forms of Chinese cabbage, chinensis and parachinensis, were by far the most commonly planted. Next were leeks (known as Chinese chives in the United States) which are considered here to be Allium odorum but are the same as the Allium tuberosum farther north. The plants look coarser here and seem to be harvested before the seedstalk much developed. Most of these are not in full bloom at harvest as was the case to the north. Many of the leeks are blanched under clay pots made especially for the purpose. The third largest area was devoted to watercress, which seems somewhat different from U.S. watercress. Also seen were taro, ginger, cabbage, arrowroot, yam, cucumber, and eggplant.

Mr. Chen Jeng-hung told us that the two Chinese cabbages are the most important summer crops. Other summer crops include the ones already listed plus water convolvulus, yardlong bean, tomato, cucumber, pickling melon, and Chinese okra (Luffa acutangula). He stated that they get an average of 12 crops per year from the same land. This is hard to understand in view of the time required for some of the crops, but some of the greens occupy the land for a very short time.

Cropping Systems for Vegetables

Essentially all the communes have interplanted one vegetable with another to make more efficient use of the land. Two crops per year is almost a minimum even in northern China and twelve per year are claimed in Canton. The most common combinations include Chinese cabbage either direct-seeded or transplanted among plants of a slower growing vegetable. If transplanted, the parachinensis type of Chinese cabbage can be harvested in about 20 days. Some examples follow with the locations given in parentheses:

- Non-heading Chinese cabbage planted at the same time between rows of the heading type, the former to be harvested first (Peking).

- Spring celery with yardlong beans interplanted later, then cabbage between the beans after celery harvested, and finally Chinese cabbage (Sian).

- Chinese cabbage among turnips, the former to be harvested first leaving all the space for turnips (Shanghai).

- Chinese cabbage among celery, to be handled as are the turnips (Shanghai).

- Cabbage planted between rows of yardlong beans, timed so that the bean plants are removed when cabbage is half grown (Shanghai).

- Cabbage with Chinese cabbage interplanted, the latter to be harvested to leave room for cabbage (Canton).

- Leeks with Chinese cabbage (both non-heading types observed). The leeks recover somewhat slowly after harvest and more than one crop of Chinese cabbage can be grown before they again use all the space (Canton).

- Cabbage at sides of beds with Chinese cabbage in the center, using the space until cabbage begins to fill it (Canton).

- Water convolvulus on the sides of beds of several vegetables, utilizing space that is too wet for most crops (various places).

- Cucumbers and yardlong beans on trellis over ditch between beds while Chinese cabbage is grown on the centers of the beds (Canton).

General Comments on Vegetable Production

Some of the notable features of vegetable production in China include the efficient use of limited amounts of land, effective water control through irrigation and drainage, and the recycling of organic waste materials. Other countries have much to learn from these practices. The increases in vegetable yields and production in the past 25 years are indeed impressive.

Nevertheless, there are several aspects in which further improvement seems possible and will probably be needed to keep pace with population increases.

Greater uniformity of maturity. While it was only possible for us to examine a few fields closely, both Chinese cabbage and cabbage seem excessively variable in maturity. Some of this probably arises from genetic variation and some from environmental sources such as disease and transplanting shock. Only research can determine the reasons, but it is certain that China has not come close to attaining the uniformity found in these crops in Japan and the United States. Consequently, the grower has the choice of losing the yield of the latest plants or delaying the planting of the next crop while the laggard plants mature.

Planting densities. A number of crops such as celery, cabbage, and Chinese cabbage are planted at much closer spacings than are used in most countries. This delays maturity, at least in cabbage, contributes to variable maturity, and results in smaller sizes in the market. We have noticed that larger vegetables are priced higher, indicating that there is a deficit of these as compared to smaller sizes. A reexamination of spacing practices might be useful.

Intercropping patterns. The clearest benefit from intercropping is in having one crop utilizing space that another crop does not need. This is clearly the situation with most of the intercropping patterns used with vegetables in China, but in some the two crops seem to be competing because both are growing rapidly at the same time. It appears in the latter case that a good deal of extra labor is needed as compared to growing the two crops separately and that the advantage is doubtful. A scientific examination of intercropping practices would be desirable.

Transplanting. We have seen a number of fields with extreme wilting of recently transplanted vegetables. The plants were rather large and time may have been lost by excessive setback as compared with using smaller plants.

These comments are not meant to be derogatory, but rather to indicate that for all its high state of development, vegetable production in China still has prospects for further improvement. There are undoubtedly other approaches as well, which would become apparent upon further observation.

Vegetable Seed Production and Germ Plasm Resources

The goal of self-reliance was repeatedly mentioned in our visits to vegetable communes and apparently they are following it by producing their own vegetable seeds insofar as possible. Certain brigades specialize in producing some seeds and supply them to other brigades in the commune. Surplus seeds and selected strains are exchanged with other communes producing vegetable seeds for areas where local germination is not possible. In Canton we heard that Professor Li Chia-wen of Shantung University had spent some time in Kwangtung Province last year in connection with a seed production project on Chinese cabbage. The seed cannot be produced well in Kwangtung Province, but can be in Shantung where it is cooler.

The limited time for discussions with vegetable specialists combined with difficulties of interpretation kept us from getting as clear a picture of vegetable seed production as we would have liked. Nevertheless, it seems to be reasonably certain that most vegetable seed is produced in the commune where it is used. This situation should lead to a great deal of genetic diversity, but we saw too few vegetables in the field at suitable stages of development to get much idea of the actual range in types. A visit to China during July would be much more informative.

Vegetables in the Peking, Shanghai, and Canton markets were not strikingly different. Cucumbers and winter squash showed the greatest diversity in fruit shape and color, the few tomatoes on display were different in size at the three markets, and taro seemed to vary from one display to another. Otherwise, there was not conspicuous variation.

The vegetable seed samples presented to us at several locations should, when grown in the United States, provide some information about the characteristics and the variation in Chinese vegetables. Until then it would be difficult to identify any specific valuable germ plasm. There might possibly be some interest in Allium tuberosum, the small Chinese leek that grows like chives. A planting can be harvested several times a year over a three to four year period, and it is widely adapted, as it is grown from Peking to Canton.

Sweet Potatoes and White Potatoes

We saw many sweet potato fields in all parts of China. We also saw small plantings on edges of fields, ditch banks, and private gardens. However, we did not see any harvesting, at least not at close range, and none in the

markets. We were told that most of them are used as food with only a few of
the roots being fed to swine. We were told that some potato research is being
done at the North China Agricultural College near Peking, but we were unable
to speak to anyone specifically concerning those crops.

We saw only a few small plantings of white potatoes, mostly in private
gardens. There may have been some recently planted fields we could not identify. Most of the production is either farther north than we went or in the
spring in the areas we visited. We did see some white potatoes in each of the
markets visited.

<u>Diseases of Vegetable Crops</u>

The following vegetable crops were either observed to be affected by the pathogens listed or were reported by researchers to be affected by these organisms. Pathogens starred are of major concern in one or more provinces visited.

<u>Host</u>	<u>Pathogen</u>
Chinese cabbage	* <u>Erwinia aroideae</u>
	<u>Erwinia carotovora</u>
	* <u>Peronospora parasitica</u>
	* <u>Brassica virus</u> -2
	<u>Alternaria brassicae</u>
	<u>Sclerotinia sclerotiorum</u>
	<u>Cercosporella brassicae</u>
Cabbage	<u>Xanthomonas campestris</u>
	<u>Phoma lingam</u>
	* <u>Peronospora parasitica</u>
Tomato	* Tobacco mosaic
	* Cucumber mosaic
	<u>Pseudomonas solanacearum</u> (S. China)
	<u>Cladosporum fulvum</u>
	<u>Phytophthora infestans</u>
	* <u>Alternaria solani</u>
	<u>Septoria lycopersici</u>
	<u>Phytophthora parasitica</u>
	<u>Xanthomonas vesicatoria</u>
	<u>Fusarium oxysporum</u> f. <u>lycopersici</u>
Eggplant	<u>Phomopsis vexans</u>
	<u>Phytophthora parasitica</u>
	<u>Verticillium albo-atrum</u>
	Cucumber mosaic
	<u>Sclerotium rolfsii</u> (observed near Canton)
Pepper	<u>Colletotrichum nigrum</u>
	<u>Gloeosporium piperatum</u>
	Cucumber mosaic
	<u>Erwinia aroideae</u>
	<u>Xanthomonas vesicatoria</u>
Cucumber	* <u>Pseudoperonospora cubensis</u>
	* <u>Erysiphe</u> **cichoracearum**
	* <u>Colletotrichum lagenarium</u>
	<u>Fusarium oxysporum</u> f. <u>niveum</u>
	<u>Mycosphaerella citrullina</u>
	<u>Pseudomonas lachrymans</u>
	Cucumber mosaic

Potato   S-Virus, (other viruses) X & Y
                              * Phytophthora infestans
                                Alternaria solani
                              * Streptomyces scabies
                                Corynebacterium sepedonicum
                                Erwinia atroseptica

Bean   Xanthomonas phaseoli
                              * Cercospora sp.
                                Colletotrichum lindemuthianum
                                Uromyces phaseoli
                                Bean mosaic

Vegetables Grown in China

common bean

yardlong bean

cowpea

soybean

hyacinth bean

broad bean

jack bean

lima bean

mung bean

pea pod

green pea

Chinese cabbage
  (heading)

Chinese cabbage
  (non-heading)

Brassica parachinensis

rape

cabbage

cauliflower

kale

brussel sprout

turnip

broccoli

kohlrabi

lettuce

stem lettuce

water convolvulus

amaranthus

malva

chenopodium

shepherd's purse

chrysanthemum

spinach

chard

beet

carrot

parsnip

parsley

celery

radish

horseradish

watercress

pachyrhiza

potato

sweet potato

yam (Dioscorea)

lotus

taro

cassava

bamboo shoot

zizania

water chestnut

ginger

onion (Allium cepa)

Chinese onion
  (A. fistulosum)

leek

Chinese leek
  (A. tuberosum)

garlic

tomato

eggplant

hot pepper

sweet pepper

physalis

sweet corn

watermelon

cucumber

muskmelon

sweet melon

pickling melon

snake melon

winter melon

snake bean
  (Tricosanthes)

winter squash

summer squash

bitter gourd
  (Momordica)

sponge gourd (Luffa)

bottle gourd

Chinese okra (Luffa)

mushroom

Vegetable Prices

Vegetable prices in Yuan/500 g. at one of about 120 markets in Shanghai.
Where range is given, higher price is for larger sizes. Two Yuan = $1.00

Leafy Vegetables	Price	Roots and Tubers	Price
Chinese cabbage (non-head)	0.25.035	taro	.06-.15
cabbage	.09	potato	.11-.18
celery	.07	garlic	.25
chrysanthemum	.08	radish	.055-.07
rape	.055	carrot	.09
mustard	.06	ginger	.44
spinach	.16	tamarix	.055
pea leaves	.23	lotus	.17-.19
leek	.12	onion	.05-.09
leek (blanched)	.28	yam (Dioscorea)	
amaranth	.055	beet	
lettuce	.065		
water convolvulus			
shepherd's purse (wild)			

Fruit Vegetables		Seeds	
tomato	.22	green soybeans	.14
snapbeans	.19	lablab	.20
yardlong beans	.11	shelled limas	.42
peppers	.11	jack bean	
eggplant	.06	Miscellaneous	
pumpkin (mooch)	.07	zizania	.23
squash (maxima)	.17	cauliflower	.19
winter melon	.045	amaranth stem	
cucumber	.09		
bitter gourd	.12		
sponge gourd	.09-.12		
snake bean			

Poultry	
live chicken	.80-.96
live duck	.90-1.43

Fruits

Fruits, fresh, stored or dried, are important components of the Chinese diet. In the northern and northeastern parts of China, apples, pears, peaches, persimmons, and grapes are cultivated, while from Nanking southward, tropical fruits predominate. According to earlier estimates, total Chinese fruit production in 1957 was around three million tons, less than 12 pounds for each person. Today, the wealth of fruits in market places suggests a marked expansion of fruit production, and this was verified by the fact that every province we visited had some research on fruits, either at special fruit breeding stations or in communes. One estimate of current fruit production is between six and nine million tons annually.

The stations where the team saw fruit research included the Fruit Research Institute of the Kirin Branch of the Academy of Agricultural and Forestry Sciences, Kung-chu-ling; the Fruit Breeding Institute, Shensi Academy of Agricultural and Forestry Sciences, Sian; the Northwest College of Agriculture, Sian (peaches only); the Institute of Agricultural Research, Nanking, Kiangsu; and the Academy of Agricultural Science, Shanghai. Research on subtropical fruits is undertaken at the Fruit Breeding Institute of the Kwangtung Academy of Agricultural Sciences, north of Canton. We did not visit that Institute but, rather, went to the Lo-kang Peoples Commune which produced a broad array of subtropical fruits with emphasis on litchi, orange, *Prunus* *mume* (for dried plums), pineapple, and banana. A total of over 30 fruits are grown on this commune.

Fruit breeding is of rather recent vintage, and the Shensi Fruit Breeding Institute dates back only to 1958. This is the largest fruit research station we visited. Breeding is largely oriented to improvement of apples and, to a lesser extent, peaches. Research on pears, grapes, and persimmon is still in the area of simple selection for improved adaptation and quality. Research on jujube and pomegranate relates to culture and local variety trials. Similarly, research on chestnut and walnut undertaken at the fruit breeding institutions is low key. We saw no research on subtropical or tropical fruits.

Kirin Province

The research at the Fruit Research Institute, Kung-chu-ling, Kirin, focuses on the development of new varieties of apples with cold hardiness, quality, and pest resistance. Standard American varieties like Red Delicious and Golden Delicious are not hardy above the 40th Parallel. Previously, only crab apples of high astringency and poor quality could be grown here. An intensive breeding program of Delicious or Ralls crossed with Malus baccata from the mountains of northern Manchuria is under way. The station has 54 hectares of fruit trees, including 160 local apple varieties, 3,000 hybrid seedlings, and a number of crab apple species. The personnel includes 14 scientists who work on apple, crab apple, apricot, pear, and grape. Pears are still restricted to the oriental hard pears based on Pyrus serotina and P. ussuriensis. There are a number of European grapes under trial, and these are grafted onto Vitis amurensis.

The station serves as an increase center for new varieties and about 10,000 apples and pears are budded for distribution to communes. Malus baccata is used as an understock for apples and P. ussuriensis for pears.

The most common diseases seen at Kung-chu-ling are scab, alternaria, and Valsa trunk canker; the latter is especially serious. Aphids are a problem, and one hybrid, #123 (Golden Delicious x Hong Tai Ping, M. prunifolia) is said to be aphid resistant. Variety #10-30 (M. baccata x Red Delicious) will keep in earthen storage for up to one year.

Some rootstock studies are under way on apples, and the Chinese are finding that transplanting M. baccata seedlings to the nursery is best done at the four-leaf stage when a 14-day physiological rest period occurs. At Kung-chu-ling this occurs in late May.

Shensi Province

The Fruit Breeding Institute of the Academy of Agricultural and Forestry Sciences of Shensi Province is located in Mei County about 110 kilometers west of Sian at the foot of the Tsingling Mountains (34°16′ Lat N - 107° 44'), at about 670 meters elevation. The average temperature is 15.9° C, with a minimum of 7.7° C in January (average). The absolute maximum in July is 42° C and absolute minimum in January is -17° C. Rainfall is 606 millimeters annually and there are 210 frost-free days. First frost occurs around November 6 and the last about April 2. The soil is a loess with a pH of 6.8.

The Institute was established in 1958 for the purpose of collecting and utilizing native fruit and nut trees from the mountains, breeding and selecting superior varieties and disseminating them to farmers, developing improved cultivation practices, pest control, propagation and pruning methods, and carrying on extension work. The Institute has 125 acres of land, works on apple, pear, persimmon, and walnut, and has some interest in apricot,

pistachio, fig, and jujube. In the southern part of the province mandarin type
oranges are cultivated.

This Institute employs 250 people including 109 scientific and technical persons although the research staff numbers less than 10. The usual
"3-in-1" relationship with the peasants is utilized. Principal research is on
apple breeding; Red Delicious x Golden Delicious cross is the basic combination. Selections are topworked onto unselected hybrid seedlings for increase
and distribution. Planting distance is also being studied. Trees are planted
in rows 3.3 m and 3.0 m in the row but this is too close (about 1000 trees/ha).
At this rate, they harvest 25 tons of Golden Delicious per year. Green manure
plus small amounts of chemical fertilizer are used. Cowpeas are planted in
late spring, disked in during September, and followed by vetch which is turned
under in spring. Golden Delicious and Red Delicious were used in many of the
cultural experiments and are performing quite well.

The Institute has a collection of 200 apples (10 popular), 180 pears
(12 popular), and 185 varieties of persimmon. Walnuts are all seedlings. In
this area persimmon is the most popular fruit and is grown individually on
farms and along the edges of fields. Concern was expressed regarding the cost
and labor required to keep such a large collection.

There are about 500 insects that attack fruits; woolly aphid is the
most serious problem. Diseases, major and minor, number about 110 of which
canker (Valsa mali) and Marsonina mali are the major ones. Canker is extremely
serious and trees are bridge-grafted here just as in Kirin to aid recovery
where canker has killed the bark on the trunk. Two viruses are reported: scar
skin on the fruits and apple leaf mosaic. The staff suggests that pear may
be the carrier of the mosaic virus as the disease is most prevalent on apple
when the two are mixed while pure apple stands show small amounts.

Persimmon research is restricted only to selection for size, maturity,
and quality of fruit. Tamopan, an old variety to us, is still popular. Varieties ripen from mid-September to early November and are now appearing on the
market.

Apples are grafted or budded onto Malus prunifolia and pears onto
Pyrus ussuriensis. Grapes are rooted from cuttings.

Shanghai Municipality

Fruit breeding at the Academy of Agricultural Science, Shanghai, is
under the Institute of Horticulture. The research covers apples and pears

adapted to warm climates, peaches, and citrus for the Shanghai area. Only in
the last few years has citrus work (mandarin orange) been attempted. Research
consists mainly of adaptation trials and assemblage of varieties. The collection includes pears (40 varieties), apples (about 15 varieties), and peaches
(60 to 70 varieties). Apple research began in 1958, but we saw few trees.
Peaches are well adapted and ripen from early June through late August. Again,
white-fleshed types predominate.

Citrus research is entirely restricted to attempts to grow mandarin
oranges and this is in its infancy. The oranges are grafted onto Poncirus
trifoliata similar to methods employed in Japan.

In general, the work does not appear to have high priority at the
Institute.

Subtropical Fruits

Taking advantage of the climate, warm temperate to subtropical, an
interesting array of fruits is grown in Kwangtung Province. This includes (in
a single commune) Prunus mume, pear, orange, litchi, banana, pineapple, and
an olive (Canarium albidum) that is eaten as a pickled fruit. Mango is grown
but it flowers at typhoon time and sets fruit poorly. Longan is also grown
and chestnuts (Castanea) on occasion. In all, about 30 fruits are cultivated,
but orange, plum (pickled), litchi (fresh and pickled), canarium (pickled),
banana, and pineapple are the major varieties. The best litchi is "glutinous
rice cake" with extremely small seeds. It is far superior here than the U.S.
variety Brewster. Some 20 varieties of litchi and 30 kinds of orange are used.
Interplanting of different fruits can be seen in some areas. Thus, on one
small mountain, pine/canarium at the top, litchi/chestnut next, then pear/litchi,
litchi/canarium, and finally at the bottom Prunus mume along with blocks of
sugarcane. A minor fruit used in season is the canombola. One characteristic
common here and elsewhere in China is that fruit is picked long before it ripens
and served quite green.

Pineapples

At the Lo-kang People's Commune in Kwangtung the Delegation saw several hectares of pineapple, one of this commune's special crops. Reportedly,
several thousand hectares are grown in southern China.

This was the Smooth Cayenne variety, planted on poor soil on a hill-
side. Part of the field had been planted with crowns in August 1974 and looked
reasonably good. In another portion, a year older, color and growth was good

(we inspected it at the end of the rainy season, probably the most favorable time) but several plants were affected by mealy bug wilt.

In a third section of the field, the plant crop harvest - generally in July and August, 22 to 23 months after planting - had been completed, and only a few scattered fruit remained. These were protected from the sun by bringing leaves up and tying them over the crownless fruit. This two-year-old field did not look particularly good; growth was uneven, and mealy bug wilt effects were evident. Fruits reportedly had been disappointingly small.

It was stated that a breeding program had been started in 1974, involving the crossing of Smooth Cayenne with other introduced varieties. In response to their inquiry, it was suggested (by Wortman) that they not pursue the long-term crossbreeding of the crop but that concentration be on improving culture of the crop, with a target of at least 40 tons of 5-pound pineapples per acre. This would require attention to plant spacing, fertilization, disease and insect control, water supply, and other factors. Mealy bug wilt requires attention immediately.

It was also suggested that a clonal selection program might be worthwhile, not only because of probable gains of 10 to 12 percent in potential productivity, but of having disease free clones with which to work in the future.

Fruit Germ Plasm in Kirin and Shensi Provinces with Potential Value for the Northern U.S.

Apple #123. This was an impressive variety in the orchard of the Fruit Research Institute of the Kirin Academy of Agricultural Science, Kung-chuling. It was selected from a cross between Golden Delicious and Red Peace (Hong Tai Ping), the latter being _Malus prunifolia_. It bears early, the third year, is annual bearing, keeps well, and as a result of its winter hardiness is used widely in northern China. The range of apples was formerly limited to 40° N latitude but this variety has extended it to 43° N. It is resistant to aphids, _Sacrotimidia mali_, which causes flower drop, and perhaps to apple scab (_Venturia inaequalis_). It is a well colored red apple with eating quality that should compare favorably with the best U.S. varieties, though fruit size is probably smaller than most U.S. varieties.

Apple #462. Another variety bred at the Fruit Research Institute, this came from a cross between _Malus baccata_ and _M. pumila_. Both parents have small fruits but this selection is larger than either parent. With a little less winter hardiness and smaller fruit size than 123, the only distinct

- 98 -

advantage of 462 seems to be earlier maturity. There may be other advantages since the staff seemed to be quite proud of this variety. Its quality seemed to be nearly as good as 123.

Apple #10-38. From Malus baccata x Red Delicious. Outstanding features are good fruit size, winter hardiness, and ability to keep for a year in pit storage.

Apple #551. Also from M. baccata x Red Delicious, this variety had the largest fruit size and was said to be very resistant to cold. It is a solid red in color, has very good flavor, and keeps well until April. They stated it showed transgressive segregation for fruit size and from what we saw we can believe that it is larger than Red Delicious.

Peach, Prunus persica x P. davidiana. This peach may be of value because it is grown by farmers in Kirin Province where hardiness is a problem with most apple varieties. We were unable to see any specimens as apples were not planted at the institute site. Apparently the cross was made at the institute some years ago but it was not clear whether the farmers are using a single clone or various seedlings from the cross. They said that Dr. Peniaczek from Poland was much interested in this peach and took scions with him. The fruit is apparently small but of good quality, ripening in mid-September. The man who seemed to be in charge of the fruit breeding was Mr. Ku Mo who seemed to be exceptionally capable, and better informed of recent U.S. developments in his field than most breeders we have met. He speaks English fairly well and is willing to send budwood to the United States at the proper season. We specifically asked that he send the hardy peach.

Apple. At the Fruit Research Institute located in Mei County, Shensi Province, Golden Delicious and Red Delicious on M. prunifolia rootstock were widely used in cultural experiments and performing extremely well. From crosses between these two varieties we saw three selections, "Glory of Yennan," "Ch'ing Kuan," and "Yen Fung." Their advantages were said to be, respectively, early maturity and better storage than Red Delicious, late maturity, and large fruit and easy management. The good performance of U.S. varieties here suggests that their varieties are worthy of trial in the United States.

Peaches. Unlike the other fruits, the responsibility for peach breeding is with the Horticulture Department of the Northwest Agricultural College, Wu-kung, Shensi Province. Professor Lu Kuang-min is director of the department. The main peach breeding accomplishment is the "Hsi-nung Winter Peach No. 18" and "Hsi-nung Winter Peach No. 19." These are said to combine the best features of their parents, one a late-bearing, and the other a sweet Chinese peach. These new varieties have high quality, late maturity, white flesh, and large fruit. Although said not to ripen until mid-October (and the fruits given to us looked very green), they proved on sampling to have good flavor and sweetness. They seem highly worthy of trial in the United States.

Ornamentals, Medicinal, and Minor Crops

Ornamentals

Despite the emphasis on production of food crops, the Chinese have an affinity for ornamental plants. In cities and rural areas, flower gardens abound around people's homes. There are numerous park plantings and street trees are used extensively. Annuals and perennials are used widely. Potted plants appear on balconies, courtyards, and around the commune homes.

Near Shanghai, there is an exceptional place called Lung-hua Nursery operated by Mr. Shu Chung-kai. This garden, established in 1954, consists of 70 hectares devoted to supplying trees, plants, and flowers to schools, public buildings, and for street planting. Over 300,000 trees are provided annually. There are three other tree nurseries in Shanghai. Those principally used in Shanghai are metasequoia, camphour, ginkgo, plane tree, deodor cedar, Chinese juniper, and our southern magnolia. Mr. Shu says the nursery has a staff of 250 employees.

In addition to tree production, a nursery of about 250 varieties of roses, 500 varieties of chrysanthemums, and about 250 types of cactus are grown for display and sale. There is also a large collection of cymbidium orchids.

But chiefly the Lung-hua Nursery is noted for its collection of dwarf potted trees (bonsai). These include specimens of trees 200 to 300 years old, as well as potted specimens of recent vintage. In addition to providing plants for display in prominent places, the nursery sends potted trees to the Canton Trade Fair for export. There are several hundred potted specimens in all.

The nursery has a training program in which several students study classical arts of drawing and painting as well as the highly detailed techniques of propagation, culture, and training of dwarf plants.

The nursery has a small botanic garden of Chinese trees and shrubs, all well labeled and set out in pleasing fashion.

Medicinal Herbs

Herbs play a major role in Chinese medicine and are sold in special shops in the major cities. In addition, each commune and production brigade has an herb garden to provide materials for use by the "barefoot doctors" in treatment of minor ailments. The basic collections of herb species are held in the botanic gardens where some taxonomic research is undertaken as well as authentication of herbs. The South China Botanic Garden at Canton, for example, has an estimated 500 species that are said to be used as herbs. These are quite different from those of north China, so the number of species used must be tremendous. One small handbook for use at the commune level contains colored plates of about 200 species of plants that can be used in medicinal concoctions.

Minor Crops

Several minor crops are usually found around the perimeters of crop fields, along roads, and in farmyard plots, all cultivated by hand. These include hemp, kenaf, castor, perilla, buckwheat, sesame, and sunflower. They may be used as sources of fiber, oil, medicinal herbs, and food. An important group of minor crops are water plants for feeding swine - such as water hyacinth, alternanthera, and _Pistia shatioides_; the latter is common in the Nanking-Shanghai area.

Pasture and Forage Crops

"China's grasslands are vast and rich." So says Hua Cheng in the January 1974 issue of _China Reconstructs_. Unfortunately, we did not see these natural grasslands and must depend on Hua Cheng and others for information concerning them.

The natural grasslands described by Hua Cheng are in the semiarid, 10 to 45 cm. rainfall belt in western China. They are composed of over 800 species, primarily Gramineae, Leguminosae, and Compositae. Hua Cheng classifies the grasslands into five groups (meadow, typical, desert, mountain, and alpine-steppe), presents pictures of each, and locates them on an outline map of China. He reports that these grasslands are being improved by using rational grazing, exterminating rodents and insects, destroying poisonous plants, drilling wells, establishing irrigated pastures, and planting "superior strains of grass." The names and sources of these "superior grasses" were not indicated.

Most of China's livestock, like the people, are in the humid eastern half of the country. In 1972 China had about 900,000 milk cows[1], 94,000,000 draft animals (buffalo, cattle, horses, mules, donkeys, and camels), 150,000,000 sheep and goats, and 260,000,000 pigs[2]; much of the feed for these animals is forage.

Milk Cows

Most of China's "black and white" milk cows are typical holsteins. They are large cows, weighing 550 to 600 kg and produce 4500 to 5600 kg of milk per year. The dual purpose shorthorn found more in the northeast produces about two-thirds as much milk. Milk cows are usually barn fed in medium to large dairies close to the cities where most of the milk is consumed. They are generally bred by artificial insemination beginning about 18 months of age, are milked an average of six years and then slaughtered. Most female offspring are kept as herd replacements; male offspring are slaughtered at an early age for veal and to conserve feed.

Milk cows are usually fed a 20 percent protein concentrate made from oil meal, grain, grain millings, brewers waste, etc., at a concentrate/milk ratio of 1:3. The Rainbow Bridge Commune on the edge of Shanghai that supplies the city with 200 tons of vegetables daily feeds its 106 cow herd much less concentrate because it wants to get the maximum amount of milk and manure from its forage. Forage at this commune consists of vegetable wastes (vines, plant parts not sold, and cull vegetables) and wild weedy grass. This weedy grass, pulled from cultivated fields or cut with hand sickles along roadsides and from land not cultivated, provides the major forage fed to milk cows. It consists of such weedy species as <u>Digitaria sanguinalis</u>, <u>Setaria viridis</u>, <u>Setaria lutescens</u>, <u>Cynodon doctyton</u>, <u>Pennisetum</u> sp. and <u>Arthaxon</u> sp., often with bits of soil attached to the roots. Rarely does the weedy grass forage contain legumes or broadleaf weeds. It is never planted. The people who harvest it and bring it to the dairies in every imaginable conveyance from hand-carried baskets to trucks are paid about 1.2 Yuan/100 kg for green grass and 5.6 Yuan/ 100 for cured hay; Shanghai prices may differ from those in other cities. Cows eat the weedy grass readily but its quality is not known.

---

[1] Weintraub, Peter, 1974. An Introduction to China's Agriculture. U.S. China Business Review, March-April, 38-41.

[2] USDA Economic Research Service publication ERS-Foreign 362, 1974. The Agricultural Situation in the People's Republic of China.

The Shanghai Dairy Company has 11 large herds that supply most of the fresh milk for the city. Dairy #2 visited September 16 was making huge stacks of weedy grass hay tied in irregular 50 kg bales. In its yard were a number of stacks of this hay totaling several hundred tons - winter feed for its 770 cow herd. The dairy also had several upright silos filled with chopped corn silage made from whole stalk corn delivered to the dairy for 2.6 Yuan/100 kg ($1.30). The sample examined had a good odor but contained no grain, suggesting that grain had been removed before the stalks were delivered. Corn stalks cut above the ear are chopped and fed immediately when in season. They may also be put in the silo.

Setaria italica millet, Vicia sativa, and perhaps other annuals are used occasionally to make hay for milk cows. In Kirin Province some prairie hay from the west is trucked to the humid east for the dairies. Alfalfa seems to be the only perennial forage planted for milk cows and very little of it was seen.

Buffalo

Buffalo grazing the weedy grasses that grow around rice paddy fields are a familiar sight in south China. They satisfy much of their forage requirement while keeping paddy walls, paths, and field borders closely "mowed." They also eat succulent water weeds in ponds where they wallow. Weedy grasses pulled from rice fields or cut from distant hillsides are also fed. The other major source of forage for these animals, particularly in winter, is rice straw. Apparently very little, if any forage is planted specifically for buffalo.

Oxen

Oxen and beef cattle require a little better quality forage than buffalo. Although they may graze along the road or in fields not cultivated, more of their forage is carried to them. Again the main forage is weedy grass pulled from the fields or cut with a sickle wherever it can be found. Wheat straw is often fed but rice straw is generally considered unsatisfactory. Females are used as draft animals but are usually worked less and fed better when nursing a small calf. Setaria millet straw is considered a preferred hay for such animals. Old oxen (about 15 years) must be sold to the government. Law forbids their owners to slaughter them.

Horses and Mules

The monogastric horses and mules cannot digest the poor quality forages that ruminants such as cattle and buffalo can handle. As a consequence, they must be fed better forage and, if working, some grain. Horses and mules are largely used to pull loads in carts along the roads. In cities like Peking and Canton they have been replaced with trucks and tractor-drawn carts, and will no doubt suffer a similar fate elsewhere as more trucks become available. Setaria millet straw, weedy grasses, prairie hay, and a little alfalfa are the principal forages fed to horses and mules.

Goats and Sheep

Goats and sheep are ruminants capable of eating a great variety of poor quality forages. They consume many broadleafed weeds and shrubs and can often be seen on their hind legs munching the lower leaves of a tree. Many peasant families in Shensi Province have white milk goats. These animals may be staked out to graze, or are fed weeds and grass pulled from the fields or cut from the hillsides and carried to them. Male offspring are slaughtered by the owner and females are grown to maturity as replacements or for sale. In the Nanking area a family sheep to supply fleece and meat takes the place of the goat in Shensi. Forages are not planted to feed these animals.

Pigs

China has four times as many pigs as does the United States. In China the pig is prized as a fertilizer factory as well as a source of food. Small pigs, usually purchased from pig farms are fed garbage, waste of all kinds, water hyacinth, and much forage until they reach marketable size. The state requires that they weigh at least 50 kg and prefers to have them weigh 100 kg. Frequently they are more than one year old when sold.

Almost every commune family owns at least one pig. Many families plant a part of their private garden plot to an annual forage legume that can be fit into crop sequence to feed their pigs. They may also use some corn or sorghum grain to fatten them.

Weeds and grass pulled from the fields or cut with a hand sickle are usually carried to pigs confined to small pens. This procedure facilitates the accumulation of the valuable manure and urine (fertilizer). It is not un-common, however, to see a pig grazing along the roadside with a rope tied to one hind leg - the other end of the rope in the owner's hand.

Freshwater Fish

China produces a substantial quantity of freshwater fish. These are usually some species of carp and are sometimes called "grass-eating" fish. In the sugarcane growing areas of south China, they are raised in small man-made ponds close to the cane. Their principal feed here consists of the lower leaves of sugarcane pulled one by one from the plant as they begin to turn yellow. The leaves bundled together may be seen floating on the water until consumed. This procedure probably cuts sugarcane yields very little. If silk is a part of the commune enterprise, the uneaten mulberry leaves, worm dung, and killed pupae taken from the cocoons go into the fish ponds. In some places pig manure from conveniently located pig barns is washed into fish ponds. Weedy grass, pulled from cultivated fields or cut from land not cultivated, is also converted to meat by these grass-eating fish. Apparently no forage is planted specifically to feed these fish.

Wild Forage in the Hills

The hills, too steep for crop, are covered with forage, shrubs, and some trees. The forages on these hills include Elymus, Bromus, Cynodon, Artemesia, Lespedeza, Melilotus, and some wild Medicago. People can be seen cutting this forage and carrying it to their livestock. Occasionally we saw animals grazing in the hills but were told that the government and communes discourage grazing because it is not favorable to conservation. From the train many of the forages in the hills of Shensi Province appeared to be bunch grasses spaced so that erosion could occur between them. On these loess soils where erosion is so intense, replacing these bunch grasses with sod formers would seem to be desirable.

We could not ascertain how well the hills beyond the reach of harvesters are being utilized. If it is not grazed or harvested, feed for livestock is being lost. It would seem that a range management system could be developed that would utilize the forage and still conserve the soil.

Pasture and Forage Research

We found little pasture and forage research in China. In the western part of Kirin Province there is a breeding program for alfalfa and sweet clover. No attempt to breed better perennial forage grasses is being made. Detailed information concerning the program was not available.

Mr. Yuan Te-ch'eng, Kiangsu Agricultural Sciences Research Institute, Nanking, together with five others is involved in the selection and breeding of superior annual forage legumes for pigs. They have collected and evaluated a few landraces of black seeded soybeans, Phaseolus aureus, and Vigna sinensis for summer, and Vicia sativa, V. villosa, Medicago hispida, and Astragalus sinicus for winter. Their objective is fast growing annual legumes that are

good pig feed, fit into crop sequences and have green manure value. They have
an active breeding program with Astragalus and report progress. It is broadcast
seeded in rice fields in October, comes on after the rice is harvested and produces green-cut, 21 to 22 percent protein, forage in April and May. They also
make silage from it for the pigs. In evaluating forage legumes for pigs they
determine yield, season of growth, and chemical composition. Quality of the
most promising selection is determined by running feeding and digestion trials
with pigs. All legumes with which they are working are cut by hand and carried
to the pigs in confinement. We plan to exchange germ plasm with Mr. Yuan.

We were unable to learn of other forage research programs that may
exist.

Turfgrasses

Turfgrasses in China are conspicuous by their absence. In Canton and
Shanghai nicely landscaped traffic circles usually include some turf. In one
such circle Zoysia matrella had been established by completely covering the soil
with irregular squares of sod. Parks, botanic gardens, hotels, and a few public
buildings in these cities usually have some areas planted to turfgrass.

Canton and Shanghai have grassed football (soccer) fields but in most
places the game is played on bare ground. China has no golf courses and at this
time cannot afford to set aside 50 hectares of her scarce food-producing land
to allow a few people to play this game.

Turfgrasses are more difficult to find in other cities. Yards inside
and outside the walls that frequently enclose rural houses are swept clean of
all grass and weeds. This is the custom symbolic of "good housekeeping" - an
attitude shared by most people in the southern United States fifty years ago.

Frequently planted turfgrasses either are not mowed or are cut (irregu-
larly) with hand sickles. During our four weeks in China we saw only one lawn
mower, a hand-pushed reel type.

China has excellent turfgrasses. Centipede grass, Eremechloa
ophiuroides, highly regarded in the southern United States, originated in southeast China, and grows wild there. The Zoysia grasses are at home in the orient.
Common bermuda grass, Cynodon dactylon, can be found almost everywhere from
Peking southward and is a major component in football field turf. Paspalum
vaginatum that helps stabilize rice paddy walls and pathways makes a beautiful
turf on low, well-watered soils.

Two sedges, Carex nigescens (French) Kreep and Carex heterolepis Binge are used for turf in the Peking area. They are reported to be more drought tolerant than grasses. These sedges, unmowed and not producing seedheads in late August, were making satisfactory turf in two plantings and very poor turf in another location in Peking.

A narrow strip of a cool season grass (believed to be Festuca pratensis) grew around the edge of a bed of salvia in front of the Kirin Academy of Agricultural Sciences in Kung-chu-ling. On a little island in a lake in Ch'ang-ch'un, considerable Poa (probably P. pratensis) mixed with weedy annuals made up the unmowed turf. Here the Poa was doing best in partial shade. These were the only cool season turfgrasses seen during our short stay in China.

The most beautiful turf we saw was growing around the base of the white marble monument in the center of the huge square in Peking. It was dark green, free of disease, and smooth although it had obviously not been mowed for some time. Close examination revealed it to be a vegetatively propagated female clone of buffalo grass, Buchloe dactyloides; a species native to the short grass prairies of the United States. Male plants of buffalo grass produce pollen-shedding heads that make the turf unsightly when not mowed. Female plants, however, bear their seeds in burrs hidden below the leaf tips, permitting them to make an attractive turf with infrequent mowing. The stoloniferous habit of buffalo grass makes it well suited to vegetative propagation. Sprigs or plants set 50 centimeters apart will completely sod a lawn in one growing season if weeds are controlled and water is adequate. In north and northeast China, west of the grasslands, buffalo grass has great potential for turf.

A Westerner can see many places in China where turfgrass could be planted: planted on athletic fields it would reduce player injuries and enhance spectator enjoyment; around rural houses it would make them more attractive and help to keep mud and dirt out of the house; it would also supply feed for a family goat, sheep, or pig that in turn could keep the lawn "mowed." Tree planted avenues in the cities could be made more beautiful by planting a strip of grass under them. Soil erosion and stream sedimentation that threatens China's river-irrigated agriculture could be reduced by planting turfgrasses on all land not devoted to growing crops and feed.

Cotton

The traditional textile fibers of China are silk and hemp. Some short-staple Gossypium herbaceum may have been grown in ancient times, but cotton did not become important in China until the introduction of American upland types. Today, cotton is the basis for a vast textile industry; the cotton crop is a major element of the Chinese economy.

We had little opportunity to study cotton firsthand and no one in the group was a specialist in the crop. Three observations from train and car windows appear valid, however: the area devoted to cotton is large; most fields are not producing very high yields; and many of the fields are highly variable in stand and plant development suggesting pronounced soil and nutritional problems, or diseases, or both. One explanation offered was the unevenness of soil following land leveling, but we were unable to study this further.

We did visit briefly with a cotton breeder at the Kiangsu Agricultural Research Institute, Nanking. One main problem results from attempts to obtain three crops per year on the same land - in an area that has traditionally grown two. One of the rotations involves interseeding cotton with barley about one month before barley harvest. There is much competition from the barley and the seedlings get off to a slow start. The cotton must be picked in time to follow with a rice crop before barley is sown again early in the year. It is a very tight, demanding sequence and requires early short-season cotton.

We were shown one field of an experimental variety that looked very good. It was derived from a G. arboreum x G. hirsutum cross. The F_1 was highly sterile as expected, but a good boll set was obtained after one backcross to G. hirsutum. Neither cytological nor genetic studies were made to determine the nature of G. arboreum chromosome elimination or the extent of genetic transfer. The line selected appeared very thrifty and rather indeterminate in growth.

In addition to use of G. arboreum, some work is being done with G. barbadense and with upland x sea island crosses. Even so, we had the distinct impression that the breeders are working with a rather narrow genetic base and are not attempting to exploit cotton genetic resources in a broad sense.

It is difficult for Americans who are not cotton specialists to appraise fields in China because they are picked frequently and few open bolls are visible at any one time. A good yield is considered to be 750 kg/ha lint and a good staple 30 mm. Cottonseed is processed in part for human consumption, some is fed to animals and some residues are used as fertilizer.

Diseases causing losses in cotton to varying degrees include:

cotton rust	potash deficiency
damping off	Rhizoctonia; Fusarium moniliforme, Pythium spp.
Alternaria leaf spot	Alternaria macrospora
Phyllosticta leaf spot	Phyllosticta gossypina
angular leaf spot	Xanthomonas malvacearum
Anthracnose	Colletotrichum gossypii
Ascochyta blight	Ascochyta gossypii
Verticillium wilt	Verticillium albo-atrum
Fusarium wilt	Fusarium oxysporum f. vasinfectum
Boll rots	Diplodia gossypina, Aspergillus niger, Rhizopus stolonifer, Phytophthora boehmeriae

The main concern in the major cotton growing provinces is Fusarium wilt. A coordinated program has been established involving about 30 scientists in 17 provinces. The coordinating institution is the Northwest College of Agriculture and Forestry (Sian), Shensi. The Department of Plant Protection has a large collection of isolates from throughout the region and they are subjecting breeding lines to inoculum in the seedling stage and also under field conditions.

In the Nanking area, Mr. Kuo, the cotton specialist at the Kiangsu Academy of Agricultural Sciences indicated that boll rots were a source of great concern to him. The most serious one is caused by Phytophthora. He is also much concerned about Fusarium wilt and is cooperating with Northwest College in Shensi.

Forestry

Although our trip did not allow for any significant review of forestry, we did meet with responsible officials in the Academy of Agricultural and Forestry Sciences in Peking. Mr. Wu Chung-lun, who studied at Yale and Duke Universities from 1945-50, was our principal contact.

Despite climate and topography well suited for forestry, only somewhere between 5 and 10 percent of the total land area, and more likely closer to the former, is forested. It also seems likely that forest cutting, especially in the commercial forests of the northeastern provinces, is proceeding much faster than are regrowth and reforestation. Consequently, unless a more vigorous program of reforestation and sound forest management are established, the forest area will soon diminish further.

Following liberation in 1949, there was a period of rapid deforestation resulting from an attempt to expand areas for cultivating food crops. In some cases, perhaps many, land that was better suited for forestry was cleared and converted to cultivated land with the concomitant worsening of erosion and silting problems.

This trend was reversed in 1958 when agricultural policy shifted toward increasing production of food on the areas then under cultivation by raising yields per acre through improved cultural practices, stimulating cropping intensity (multiple cropping), and improving the utilization of water resources. At the same time a national campaign was organized to beautify the countryside. Communes were encouraged to plant trees for shelterbelts, soil conservation, and forest production on a continually renewable basis.

Many millions of trees have been planted during the past 15 years in shelterbelts, along roads, railroads, and streets throughout the country. This aspect of the afforestation program has been highly successful.

The more difficult problem of reforesting mountainous lands that are poorly suited to agriculture has been less successful. Although we traveled long distances in China by train, we saw only a very modest number of sizable re-forestation projects. On many of these stand establishment was poor, often less than 25 percent survival. The one exception was the rather large pine (P. massoniana) plantations on the rolling hills in Anhwei Province, northwest of Nanking, where stand establishment was excellent.

In the reforestation projects we observed on denuded mountainous areas, two species of pine (Pinus massoniana and P. tabulaeformis) had been used extensively. Neither of these species were impressive from a timber production standpoint. Both are generally twisted and crooked, although we were told that some seed sources of both species have been found which possess better form. Some proveniences (seed sources) of these species have been reported to be better suited to the re-establishment of forests on thin, eroded rocky mountainous soils, a characteristic that is of great importance on many eroded sites in mountainous areas. Another problem that must be overcome in order to re-establish the full benefits of forest cover and watershed protection is the collection of needles and twigs from forest plantations. Many forested areas near cities are literally swept bare of humus.

The plantings of P. massoniana, which are more widely used in the south, were heavily damaged by an insect that affected the terminal shoot, further contributing to poor form development.

In the northeast and north China plain areas, which are largely agricultural, the afforestation projects observed by the team consisted mainly of shelterbelts along the edges of fields and roads. For these uses emphasis had been on species of populus, including P. mafemowiczii, P. simonii, P. suaveolens and P. nigra. Originally, P. canadensis was planted but is apparently susceptible to diseases - both rust and a leaf spot - and its planting has been largely discontinued. We were told, however, that successful reforestation projects are being carried out in the mountainous areas of Kirin, Heilungkiang and Liaoning employing principally the indigenous species Pinus koraiensis, which is one of the most valuable timber species of northeast China.

In addition, there is great reliance on willow (Salix matsudana) and we saw more of these trees (poplar and willow) than any other in north China, used either separately or in combination. They are also the main trees for planting along roadsides.

Although several species of Eucalyptus, including E. citridora, E. excerta, E. rudus, and E. teretieornus grow in parks, along roadsides and railroads, we saw no extensive plantings of any of them, and most that we saw growing were of poor form. It would seem that the potential use of Eucalyptus in southern China might justify further introductions and experimentation.

In the south, there is some attempt to use exotic species, of which Cunninghamia lanceolata is the most promising. In warm regions of the south, Phoebe (lauraceae) is used extensively for afforestation. Wood for construction purposes is scarce and there is a great reliance on concrete. Railroad ties are largely of concrete as are telephone poles.

Bamboos remain important both for conservation in the south, as well as for poles and furniture. Phyllostachys bambusoides, and P. pubescens are the main timber types and will continue to be of importance for years to come. For example, in a large generator plant, we observed thin bamboo strips being used in large quantities as insulating material in generators of all sizes.

There is little interest in chestnut as a forest tree. It is used exclusively for nut production. However, it was mentioned that Castanea henryi is a better forest type than C. mollissima. Seed of this species was requested.

Metasequoia glyptostroboides is making its appearance in large numbers in all of the areas visited, often in the most unlikely sites. It is produced both from seed and cuttings. Seed still comes from the original stand discovered in 1945. Taxodium distichum was observed growing well in both Shanghai and Canton.

Quercus is of interest. About 10 years ago Q. acutissima was being used but now emphasis is on Q. variabilis because it is faster growing.

Among introduced species that may play an important role in future reforestation and soil conservation programs of China is black locust (Robinia pseudoacacia). This American species has been planted along roads, railroads, and some mountainous slopes as well as along city streets. It was observed to be reproducing well everywhere, including many difficult sites on steep slopes with shallow soils. We were informed that there is now a growing interest in this species for use in reforestation projects where control of soil erosion is a prime consideration. The plantings that we were able to review were free of locust borer, the worst insect pest of black locust in the United States.

Forest Research

We saw no evidence of any significant forest research effort, regarding timber production, nor were we able to learn of any from conversations with the few foresters we met. However, there is a forestry institute with 8 to 10 scientists located in the College of Agriculture, Sian, Shensi.

We did not see any experimental forest plantations where different forest species or seed of different proveniences of the same species were being studied for adaptation, survival, growth rate, tree form, and resistance to diseases and insect pests. Neither did we see or hear of any sizable forest genetic or treebreeding programs. Nor did we see any of the large reforestation programs that have been reported, though we often traveled by train. When we inquired about the success of the extensive aerial seeding operations that had been reported in the United States, we got the impression that they had rarely been successful. Moreover, we were told that the areas that were seeded from the air were small.

It would appear that up to the present time the major government emphasis has been the increase and stabilization of food production; only minor attention has been given to reforestation and soil conservation in the mountainous headwaters of its principal rivers, upon which its irrigation and much of its food production potential depends.

It is very possible that the time may soon arrive when the government will launch a major drive to reduce erosion and silting in the Huang Ho (Yellow) River basin. Many of the main tributaries of this large river system headwater in the mountains and cut their way through the extensive, highly erodable, loess deposits of the provinces of Kansu, Shensi, Nigsiahua, and Shansi. This gives rise to the heavy silting that rapidly fill reservoirs with mud, fouls irrigation canals and drainage ditches, and endangers flooding on the lower reaches of the river. There is little doubt but that the silting and erosion on the Yellow River system remains one of the greatest unsolved problems of Chinese agriculture.

If and when this problem receives priority it will probably be handled quite effectively. The government has the political structure, manpower, and discipline to launch a massive campaign if it develops the necessary technical background information. We learned that the hydraulic engineers of the Water Resources Ministry have established experiment stations on some watersheds of the tributaries of the Yellow River. We could not learn if this is a broad interdisciplinary research program embracing the development of effective engineering devices, check dams, etc., and extensive evaluation of the suitability of plant species - including forest, shrub, and grass species. The development of such data now will determine how rapidly a successful reforestation and soil conservation program can be implemented once the decision is made to encourage it.

Urban Forestry

China is undertaking a remarkable campaign to "green-up" cities, industrial complexes, and highways. Many city trees are Platanus accrifolia (London Plane), and are grown in a headed fashion with the leader removed to force the trees to branch out forming a canopy over the highway. In some areas of Nanking, the divided highway is planted six rows deep.

Cedrus deodara has made a marked appearance on the streets of Nanking and Shanghai. These are alternated with Juniperus chinensis and Metasequoia. Other trees commonly seen include Salix babylonica, Magnolia grandiflora (grafted onto M. denudata), Liriodendron chinensis, Liquidambar formosana, Pinus thunbergii, P. densiflora, and P. bungeana, the latter mostly in old temple areas.

Ulmus pumila is used in northern dry regions and Zelkova sinica in warmer places around Nanking. Farther south, broadleaved evergreens appear and Cinnamomum camphora is used extensively in new industrial suburbs of Shanghai.

Trees are produced in city-owned nurseries (three in Shanghai), at experimental stations, and commune headquarters.

In general, one can conclude that along with other urban and rural life improvement programs, tree planting for conservation and aesthetic purposes is by no means neglected. Other than Magnolia grandiflora and a few species reflecting Japanese influence, parks and memorials are planted exclusively to native trees and shrubs, most of which are common exotics in the United States. Box elder (<u>Acer</u> <u>nigundo</u>) originally introduced for shelterbelts, is commonly grown in northeastern China, reflecting some influence of American travelers to China decades ago.

Cropping Systems

China has made remarkable progress in multiple cropping. Building upon a long tradition of keeping a crop on the land throughout the growing season, the Chinese have markedly increased land utilization in recent years. Through water conservancy projects they have increased control of water, providing drainage during periods of excess water and irrigation during dry periods. Also, plant breeding efforts and crop culture practices have been oriented toward obtaining the maximum number of crops possible. This has resulted in the world's most concentrated use of land resources during crop growing periods.

The following notes summarize our impressions of cropping systems (where vegetables are relatively minor - the vegetable crops section of this report covers cropping systems involving these crops).

1. As viewed from the train from Hong Kong to Canton, all tillable land was in use, the lowlands for paddy rice, intermediate areas for string beans, cassava, sweet potatoes, sugar cane and ramie, some lotus and water chestnuts. Rice was only recently planted. Hillsides were being reforested.

There was some evidence of other vegetables planted under beans (cowpeas) before the latter was harvested. This was seen again the next day at a commune field test site where vegetables were planted under unharvested cowpeas.

2. Visiting Nan-yuan People's Commune we were told that a total of 80 varieties of vegetables were cultivated at this commune (it grows a little more than 1000 hectares of vegetables annually). The main vegetables were Chinese cabbage, cabbage, cucumber, peppers, eggplant, tomatoes, leeks, and beans.

Chinese cabbage was grown on one plot with a similar lower growing plant in between.

3. A report from CAAFS indicated that following the cultural revolution the practice of growing only one crop of rice was changed in the central and northern part of China as follows:

- Up to 29° N latitude: double cropping of rice.

- North of the Yellow River: one crop of wheat followed by one of rice (important system of north China).

- 114 -

- In the Yangtze River basin: one crop of rice, one of alternate crop on unirrigated land; two crops of rice where irrigated.

These advances were made possible by two changes:

- So-called capital construction - water conservancy projects which increase irrigation and improve drainage and water control generally.

- Improved varieties (including shortened growing season for rice) for both crops.

We were shown rice plots that had followed winter wheat and were told most of the rice growing on nearby communes was following wheat.

4. Kirin Province. Three kinds of cropping systems involving more than one crop were observed. First, double cropping of wheat and corn where wheat was planted in early April and corn about three weeks later. The wheat was harvested in July and the corn was nearing maturity and looked good, although somewhat shorter than surrounding plots. A row spacing of 60 to 75 cm appeared best. We saw intercropping of corn and soybeans with width of strips from 2 to 6 rows. There was some confusion as to which width was best, although the greater emphasis on corn in this area dictated much more corn than soybeans in the patterns.

In addition to corn-soybean combinations, we saw corn-millet, sorghum-millet, sorghum-soybeans, and one field of two sorghums, one tall the other short. From the air flying south from Chang Chun to Peking we judged that at least half of the corn and sorghum were intercropped.

There was some mixed cropping, that is, two crops planted in the same row. This was much less extensive, however, than intercropping in alternate rows.

5. Yellow River Valley. The crops we viewed were probably the second crop in a double cropping sequence, winter wheat or some other winter or early spring crop having preceded the ones now in the field.

There was little intercropping in this area, quite in contrast to fields in higher rainfall and lower temperature areas in the northeast. Intercropping is generally of no advantage when rainfall deficiency seriously limits crop growth. There were a few fields with intercropping combinations such as corn-soybeans, and corn-sweet potatoes, but solid plantings of corn, cotton, millet, sweet potatoes, and sesame dominated. Fruit trees were always intercropped with soybeans or vegetables.

6. Shensi Province. Here we noted the degree of local control over the use of intercropping. One county, northeast of Sian, about 100 km to the

Shensi Academy of Agricultural and Forestry Sciences, essentially had no intercropping. By contrast, 90 km straight west of Sian at the Fruit Research
Institute we noted an unusual assortment of intercropping patterns fully as innovative as we had seen in Kirin Province. There appears to be no biological
reason for this difference, suggesting that it is likely local custom.

   7. Five-Star Brigade, Red Flag Commune, near Sian. Since the commune
was organized the number of crops grown per year has increased twofold, three-
fold or even fourfold in some cases. They gave an example where celery, the
first crop, was interplanted before harvest with kidney beans. This was followed
by cabbage and finally Chinese cabbage. Apparently, there was some relay planting
in the case of the first three crops. A total annual yield of 23,000 chin per mow
(172.5 MT/ha or 77.5 tons per acre) was obtained from these four vegetable crops.

   8. Kiangsu Province. Two basic cropping patterns are followed in
Kiangsu Province. In the south, winter wheat, or sometimes winter barley, is
followed by an indica rice variety and then a second crop of japonica rice that
is photoperiod sensitive. In the northern part of the province, winter wheat is
followed by rice and a green manure crop which is then turned under. Winter
wheat follows, starting the cycle again. In the north they also use wheat,
sweet potato, and winter fallow followed by corn and then wheat which starts the
cycle again.

   9. Shanghai. Two basic grain cropping systems were described for the
Shanghai area; both involve winter grains (wheat or barley). In the first, winter
wheat or barley is planted in early November and harvested in the last half of
May. An indica crop of rice is transplanted immediately and is ready for harvest
in late July or early August when a second crop of japonica rice is transplanted.
This in turn is harvested in late October or early November and the cycle starts
again.

   The second system involves only two crops, wheat (or barley) and cotton,
since the growing season for cotton is longer than for rice.

   In the vegetable growing area around Shanghai there are up to four
full crops a year, as there is little time between the harvest of one crop and
the planting of another. As in the case of rice, transplants are commonly used
to reduce the period between planting and harvesting. Hand watering of individual
plants helps assure success of the transplanted crop.

   10. Kwangtung Province. This province has several interesting cropping systems. Those involving rice are not greatly different from the systems in

use in irrigated areas of the Yangtze River Basin. Early rice is sown in the first half of March, transplanted in early April, and harvested in the last half of July. A second rice crop is sown in June, transplanted in late July and harvested in mid-November. A winter crop, vegetable or grain, is grown between November and March. Rice areas are rotated with areas on which vegetables are grown.

A most interesting cropping system is found south and southeast of Canton. Much of this area is low-lying and in the long distant past was likely subject to periodic flooding. Fish ponds have been dug and the soil removed in the process provides upland for other crops including sugar cane, mulberry, banana, soybeans, peanuts, and vegetables. After the mulberry leaves are stripped, the twigs are cut and cabbage or lettuce is interplanted. Refuse from the crops, including the residue from silk worm growth and cocooning are returned to the ponds. Essentially every square meter of land is being utilized.

THE BIOLOGICAL SCIENCES IN CHINA

One objective of the U.S. Plant Sciences Delegation was to determine the status of science fields related to biology and agriculture. Several impressions emerged from visits to central and provincial academies and institutes, discussions with scientists and administrators, and observations of the quality of farming and the nature of experimental work at the production brigade level in the communes. Some notion of the general status of scientific work is revealed in the discussions of individual crops and in the overview portions of this report. Also, our hosts in China repeatedly asked for suggestions for improving their work and near the end of the month's study the group prepared these. (Appendix 7.) The suggestions obviously reflect the U.S. team's impressions, including the following:

1. China has been remarkably successful in farm application of the accumulated knowledge of generations of peasants' experience as well as past scientific research. Production on all farm plots in any locality appears to be reaching the highest level of output permitted by such accumulated local knowledge. As a result, there is little variation in villages of the quality of farming.

To accomplish this, China has required that most scientists devote a major portion of their time to the nation's production efforts, working closely with the technicians and farmers. Most scientists are expected to spend one year out of three in residence at communes, and to participate actively as consultants to communes when back at their research posts. This undoubtedly has made most scientists highly production-conscious, given agricultural scientists a knowledge of farming and farm problems, and broken the "elitism" and "urbanism" discussed at some length by Dr. Stavis in his Cornell University monograph "Making Green Revolution." It all seems to have made important contributions to the rapid gains in output on tens of thousands of communes that were greatly needed. Everyone had to be involved to get grassroots progress on such a scale in such a short time span.

2. China probably has exploited much of the accumulated knowledge
which could more quickly be put to use, with or without difficulty. However,
if the nation is to continue to improve agricultural yields and production, a
dynamic, production-oriented and much more sophisticated fundamental research effort will be required. Chinese agricultural and biological science does not
seem to be as well supported as needs will require. In fact, it appears that
during the last several years little of the needed fundamental agricultural research has been done because of attention to immediate production problems, lack
of conviction of central authorities that more fundamental research is important,
lack of interest by scientists in problem oriented research, or more likely a
combination of such factors. One gets the impression that much Chinese agricultural research is currently stagnant, as the research establishment undergoes
change. The U.S. group was told that Chinese authorities are aware that the
push for production has had such effects and with the completion of reorganization there will be support for an all-out research effort; one in which the full
creativity of Chinese science will be brought to bear on agricultural development.

3. There is evidence of the lack of support for science: poorly
equipped laboratories, especially in academies and institutes of agricultural
sciences; little laboratory research activity in all institutions visited
(although some institutes of Academia Sinica reported results of recent research);
and a noticeable lack of field experiments other than variety tests or plant
breeding plots - at most institutes or at brigades being assisted by institutes.
Progress achieved so far has been in spite of the lack of facilities and a
dynamic, more basic research effort. But progress is not likely to continue over
the long term unless this weakness is corrected.

4. In general scientists are not aware of work in other provinces.
One wonders to what extent scientific cooperation and coordination on similar
problems in different provinces is encouraged. It is well understood that local
authorities might take a dim view of work not directly focused on their own
problems, or of scientists traveling outside their assigned areas; this is the
case in many countries. But China will need to achieve a united, strong national
effort on research of national or regional importance. Present efforts are not
nearly as effective as they could and should be. On the other hand, few nations
have achieved the level of coordination of scientific work from which China could
benefit.

A question then arises: Would the U.S. delegation advocate a return to "elitism"? Certainly, China must develop for each crop or major problem a critical mass of highly talented research people - a group to deal with the more difficult problems that require the best minds, with the best equipment and facilities, solely responsible for the search for solutions. Such groups could be developed either at national academies or provincial ones, provided they have the facilities and support needed and the ability to work nationally. These groups could be of sufficient size so that one-third could at any time be rotated out to provinces, making room for provincial people to come in for advanced experience. These teams should be expected to set new and higher standards of work for colleagues elsewhere in China, and to serve the entire country by leading the national attack on major agricultural problems. They should also have the responsibility of knowing about relevant work going on elsewhere in the world and of quickly incorporating such advances into the framework of Chinese agriculture. But the teams should not be permitted to lose their contact with China's farm people and their problems, nor to lose their present concern for increased productivity. They must continue to serve the people as they now do.

5. China's scientists have been, and still are, out of contact with the mainstream of international scientific activity, again probably because all national efforts are focused on production. Consequently, we found scientists generally unaware of advances occurring elsewhere, many that could be of great benefit to Chinese agriculture.

Agronomy and Soils

China's farmers are skilled and dedicated people and the quality of their farming reflects this. In fact, one could almost say that there is no "farming" of crops in China, that even the cereal crops are "gardened." Therefore, the nature of the agronomic research required to move the country to still higher yield levels will have to be highly imaginative, based on the latest advances in science, and be carefully done and widely tested where indicated. Such work was not seen. For example, we did not see work on plant populations as they interact with timing and amounts of applications of water, on specific nutrient elements, or on aspects of plant architecture as they affect culture of the crops. About the only type of field experiments we did see were a few maximum yield trials, which of course are quite important. It must be noted though, that we visited only a few research institutions and that the group specifically had requested opportunities to see plant sciences work, not that in soils. Nevertheless, the absence of active field and laboratory work on some of the major agronomic problems was noticeable at all institutions visited.

Botany and Botanical Gardens

Our visits included the Institute of Botany, Academy of Sciences, Peking, and the Department of Botany, Sun Yat-sen University, Canton.

The Peking institute is housed in a former temple complex and is eventually scheduled to move to the Peking Botanic Garden which is presently closed to visitors. The institute has 300 scientists and technicians working in taxonomy, geography, paleobotany, morphology, cytology, and laboratories for cell fusion, metabolism, hormones, and photosynthesis.

This section covers only floristics and related subjects. (Others are dealt with elsewhere in the report.) The major effort is the coordination of the long-term preparation of the flora of China - eventually to run between 70 and 80 volumes. Work is assigned to botanists around China who will cover families for which they are specialists. No less than six artists at Peking are engaged in preparing line drawings for the flora. The institute has the largest herbarium in China but because of the temporary quarters it is somewhat scattered. However, the institute is quite prepared to receive and loan study material.

There is an Iconographia Cormophytum Sinicorum partially completed (vol. I, 1971; vol. II, 1972); volume III is complete but yet to be printed. The first two volumes cover ferns, gymnosperms and dicota through Cornaceae. The units of phytogeography and ecology are developing new vegetation maps of China and conducting studies on plant indicators of soil types with particular reference to crop potentials, and growth patterns of selected crop and wild plants on acid, neutral and alkaline soils.

The biology department of the Sun Yat-sen University covers biochemistry, entomology, herbal plants and taxonomy, and zoology. There is a fair herbarium that team members were invited to examine for their special interests and several interesting families were seen. A good collection of bamboos is cultivated on the campus (formerly Lingnan University), and was initially started by Dr. Floyd McClure. We saw the clump of Lingnania, a genus of bamboo established by Dr. McClure.

In general, classical taxonomy directed toward the production needs of the state seems to require a justifiable role. The aspects of floristics as related to agricultural and forestry practices and herbal botany seem to have high priority. A close association has been developed with pharmaceutical factories and herb-producing communes.

As with classical botany, botanical gardens are being reevaluated in light of the production orientation of Chinese agricultural society. As a result, it was possible to visit only one botanical garden, the Kwangtung Botanical Garden, Canton. It is headed by Professor Ch'en Feng-hua who started the famous botanical garden at Lushan as well as others. It appears that until botanical gardens have been accorded a precise role in the new order, their leadership and status is somewhat vague. They are operated jointly by the Academy of Science and the provincial government where they are located. Most do not have current seed lists but Lushan will have one, perhaps this year. Certainly botanical gardens will play a more active role in economic botany and education than in the period prior to the Cultural Revolution (1966).

The Kwangtung Botanical Garden at Canton (1959) is on former waste land. It has some 300 hectares of land, of which about one quarter is planted. It has only subtropical and warm-temperate elements totaling about 3000 specimens. Presently, the garden is closed to the public except for special arrangements. To emphasize the shift in direction, priority for public development is the fern collection, now about 50 species, displayed in an ornamental manner, a collection of cymbidium orchids, water plants and other ornamentals. A small pavilion overlooks a fine planting of Taxodium, Cryptomeria, Metasequoia, Glyptostrobus, and Araucaria. Finally, an herb garden is now being developed and contains an estimated 500 species of herbs used in south China. Work includes identification, culture, use, and preparation. Because herbs play a major role in Chinese medicine, are used on every commune and are given equal status with western medicine, it can be readily understood that a logical and practical role for botanical gardens of China will relate to herbs. In botanical institutes such matters as classification, the nature of activity, and remedial qualities are studied. Handbooks for the use and gathering of herbs by communes have been prepared for different regions of China and these are usually illustrated with colored plates.

Paleobotany Department, Institute of Botany, Academia Sinica, Peking

This is a very small department headed by Mr. Jen Hsu. He has studied in Sweden and worked some nine years in India. He met Ralph Chaney on the latter's visits to China, and, although he has not been to the U.S. is well known to paleobotanists there. Mr. Jen is primarily interested in fossils of the Paleozoic period and China is very rich in beautifully preserved fossil materials of this time range. He has done some archaeobotanical identification although that is not his specialty. It was his impression that sorghum occurred in Yang-shao sites, but the material was returned to the archaeologists. The work of the department supports the geological survey of China and should be of considerable help in locating mineral and petroleum deposits. Some recent finds of cones of _Picea wilsonii_ Mast. near Sian and the Pan-p'o area indicate the climate there was considerably cooler 10,000 to 20,000 years ago. This is in no way surprising but it is good to have the fossil evidence.

Herb Plants Growing in the Kwangtung Botanical Garden, Canton, China

Abacopteris	Callicarpa	Eomecon	Morinda	Sambucus
Achyranthes	Calophyllum	Epothylobus	Nandina	Sanguisorbia
Acorus	Catharanthus	Eryngium	Ocimum	Saururus
Actinidia	Chloranthus	Euonymus	Ophiopogon	Saxifraga
Adena	Circium	Euphorbia	Orixa	Sedum
Agapanthus	Cissus	Evodia	Orthodon	Selaginella
Agrimonia	Citrus	Gynura	Phytolacca	Senecio
Ainsliaea	Clinacanthus	Hosta	Pimpinella	Solanum
Akebia	Coix	Ilex	Piper	Spilanthes
Aloe	Cordyline	Ipomoea	Plantago	Stauntonia
Ampelopsis	Crinum	Jasminum	Platycodon	Stephania
Andrographis	Croton	Justicia	Plectranthus	Strobilanthes
Anetum	Cryptolepis	Kadsura	Plumbago	Symphytum
Anemone	Curcuma	Kaempferia	Polyalthia	Tacca
Angelica	Cyathea	Ligustrum	Polygonum	Talinum
Ardisia	Cynanchum	Lindera	Psychotria	Thalictrum
Aristolochia	Damnacanthus	Litsea	Pteris	Tylophora
Artemesia	Desmodium	Lonicera	Quisqualis	Viola
Asarum	Dicliptera	Loropetalum	Rauwolfia	Wrightia
Aster	Dracaena	Lysimachia	Rhamnus	Zanthoxylum
Atalantia	Drymia	Macleaya	Rhinacanthus	Zizyphus
Baphicacanthes	Dysosma	Mahonia	Rubia	
Begonia	Elephantopus	Melothria	Rumex	
Berchemia	Emilia	Mentha	Ruta	
Botryopleuron	Entada	Mirabilis	Salvia	

Genetics and Plant Breeding

Interest in genetics in the People's Republic of China is limited almost exclusively to those segments which offer promise of direct application in plant and animal breeding. This restricted viewpoint applies equally in both teaching and research. The agency having primary responsibility for genetic developments is the Institute of Genetics under the National Academy of Sciences. This institute was established in 1959 for the study of variation in plants and animals.

Crop improvement research is conducted by a large number of provincial branches of the Academy of Agricultural and Forestry Sciences, agricultural colleges, and the genetics sections of universities. No attempt will be made to separately review work at each of the institutions visited. Instead we shall use the subdivisions of the Institute of Genetics as major headings to record observations and generalizations.

Research in genetics and breeding may be classified under four separate headings: Industrial and Medical Molecular Genetics, Plant Genetics, Animal Genetics, and Heterosis.

Industrial and Medical Molecular Genetics

The organism of primary interest is <u>Bacillus</u> <u>subtilis</u>. A product has been obtained from cultures of this bacteria which is highly toxic to the larvae of Lepidoptera. Extensive commercial production of the substance is under way although the chemical nature of the compound has not been established nor has it been named.

Studies are under way to isolate strains of this bacteria for specialized enzyme production. Amylase is already being produced on a commercial scale. Both transduction and transformation are being used in a search for strains having high antibiotic potential. Here as in all other institutes a third or more of the staff are regularly assigned to work with production brigades at either a factory or agricultural commune.

Plant Genetics

Crop improvement work follows both traditional lines and exploration of new techniques. Two different techniques are being used to develop fertile, true-breeding sporophytes from gametes. The first involves pollen culture of anthers taken from F_1 plants. At least four institutions are working on this problem: the Institutes of Botany and Genetics under the Chinese Academy of Sciences of Peking, the Academy of Agricultural Sciences, and the Institute of Plant Physiology, the two latter operating under the Shanghai municipality.

The general approach was similar for all; the first two concentrating on wheat and the second two on rice. Anthers with pollen at the early to mid-mononucleate stage are seeded on agar containing the standard Murashige and Skoog medium supplemented with additional amounts of sucrose, thiamine chloride, auxins, lactalbumen hydrolysate and 2,4-D. The developing calli are transferred when about 1-3 mm. in diameter. M-S medium is used without 2,4-D but containing 2 mg./liter of IAA and kinetin. Roots or shoots may become visible in about 20 days. Rather sizable numbers of calli produce only albino plantlets.

When plants are 0.5 to 1.0 cm. in height they are transferred to **White's medium with reduced sucrose (1.5 percent) content, supplemented with 1 mg./** liter IAA. After further growth the young plants are transplanted to pots and grown to maturity. The mature plants may be sterile, have fertile sectors or may be completely fertile. The seeds produced give rise to fully fertile and uniform progeny.

This standardized procedure has definite limitations as a plant breeding technique. Methods have not yet been developed to overcome a very marked varietal effect. Calli can be obtained from some genotypes with reasonable success while other genotypes are almost completely intractable.

At the Institute of Plant Physiology, Shanghai, initial studies have been directed toward protoplast fusion, in part as a tool for studying developmental morphology but also as a potential plant breeding technique. Major work thus far has dealt with the lysing of the cell walls of different tissues (leaf and root) of many plants: barley, wheat, corn, rice, tobacco, pepper, eggplant, peas, soybeans, morning glory and petunia. Protoplast fusion has been found to occur spontaneously as judged by the appearance of protoplasts with two nuclei. Cell division and regeneration of cell walls from barley and carrot protoplasts has also been observed.

Fusion has been observed between tobacco and carrot and between Vicia faba and barley protoplasts. No calli have yet been developed from these fusion products.

The second method being used to obtain homozygote sporophytes involves wheat. Florets of _Triticum aestivum_ are emasculated. Some 7 to 10 days later the emasculated florets are pollinated by _T. durum._ Due to the delay in pollination seed set is low. Hybrid progeny can be identified and are discarded. The _T. aestivum_ progeny are presumed to have resulted from the stimulation of an egg cell. A number of fertile sporophytes have been produced by this method.

There seemed to be little appreciation that with gametic as with zygotic sampling, the most probable outcome with small numbers is something approximating the mean of the parental types. Only as sample size becomes reasonably large can one hope to recover new genotypes which provide the desired combination of genes controlling yield ability, quality, and disease and insect resistance required for real progress.

The choice of parents for conventional breeding or for anther culture in both rice and wheat could be materially improved by China's participation in one or more of the several international uniform nurseries. Participation would provide knowledge of and access to the best sources of disease and insect resistance thus far identified throughout the world.

Traditional techniques are being used in the improvement of wheat, rice, millet, rape, cotton, and the many vegetable crops. As new varieties become widely grown susceptibilities to either diseases or insect pests soon become evident. While the important pests have been identified, only limited work has been done in identifying sources of resistance and even less on the inheritance pattern of such resistance. Such studies are essential for the efficient use of resistant types in a breeding program.

Improvement of the vegetable crops has possibly progressed at a slower rate than with the cereal, fiber or oilseed crops. This is due to the great variety of vegetables grown.

Animal Genetics

We did not see any of the work in animal breeding conducted by the Institute of Genetics. We were told that considerable breed improvement had been effected in both cattle and swine. A small white milk goat reportedly has been developed and is gaining acceptance in Shensi Province and other northwest areas.

Heterosis

The work on heterosis is largely confined to the development of
inbred lines of corn and sorghum and their evaluation in hybrid combinations.
American and Russian corn lines are being used in addition to those developed
locally. The genetic-cytoplasmic sterility system developed for sorghum in
the United States is being used at several locations in the breeding programs.

Corn is widely grown throughout China. In many areas this crop
occupies the land for only a portion of the season; only it is a full-season
crop in the northeast. For this reason differences in maturities are much
less than would be encountered in other areas of the world having a similar
latitudinal range. Several U.S. lines are used throughout this range. These
lines were developed in the late 1930's and early 1940's (Oh43, C103, R181,
W20, W24, etc.). Most of the hybrids which have been "popularized" represent
combinations of locally developed and introduced lines. Such combinations
would be expected to exhibit the highest levels of heterosis due to genetic
diversity.

The use of a relatively small group of U.S. lines over a large
area suggests that progress might be greatly accelerated by: the introduction
of a new series of lines having higher combining ability and greater disease
and insect resistance; combining these new lines in some systematic fashion
with the locally developed lines in commercial use; and evaluating the single-
cross hybrids in a series of cooperative tests distributed throughout the area
where corn is an important crop.

The generally high level of heterosis between local and introduced
lines provides an excellent opportunity for the establishment of a recurrent
selection program. Some consideration is being given to this possibility in
the corn program at the Kirin branch of the Academy of Agricultural and
Forestry Sciences.

The genetic-cytoplasmic sterility system in sorghum, introduced
from the United States by way of Africa, was used at the Atomic Energy
Utilization Research Institute of the CAAFS, Peking. Two hybrids, Nos. 10
and 12, reportedly were developed. The cytoplasmic sterile source (A line)
has also been used at other locations in hybrid development and production.
Most of the hybrids currently grown are tall but interest in dwarf types may
increase due to the need for certain intercropping systems.

The native landraces of kaoliang exhibit considerable variation in seed color. All of the hybrids used commercially, however, are brown seeded. All brown seeded sorghums are high in tannins. Nutrition studies in the United States have shown that tannin interferes with protein assimilation and the PER values of such types are always low. A shift to white, yellow, or red seeded types could be easily effected and would result in improved nutritional value.

Little recognition has been given to the fact that any effective system of breeding leads to an increase in genetic uniformity and therefore greater potential genetic vulnerability. This risk can be minimized only by adequate germ plasm collections which must be the source of resistance to any new races of pathogens or insect biotypes. Such collections do not now exist in China.

Horticulture

It is not possible to give any overall description or evaluation of research on vegetables based on our visit. When we visited a research center, most of the time was needed for group discussion of the major crops and even though one or more vegetable specialists might have been present, there was little time to talk with them. The time of our visit was such that there were few experiments in the field, and weather, lack of time, or distance to plots usually prevented us from seeing field experiments anyway. Consequently, our information and impressions are inadequate and fragmentary.

In Kirin Province we spent two days at the Academy of Agricultural Sciences and found no work on vegetables. Only at the end of the visit and after several inquiries about vegetable research in the province did we learn that there is a Vegetable Research Institute in Ch'ang-ch'un. It was formerly a part of the academy but now is under the municipality. It was said to have 50 to 60 workers.

In Peking, we had a two-hour discussion with vegetable specialists from the Academy of Agricultural and Forestry Sciences, the Institute of Genetics of the Academy of Science, and the Provincial Institute of Agricultural Science, but had no chance to see any of their field experiments. Major topics and discussions follow.

1. Disease resistance in tomato. <u>Fusarium</u> wilt was said to be a serious problem only in the south of China but questions were asked about how to test and select for resistance. There was much greater interest in TMV resistance and there was a long discussion about the various genes for resistance, of which the Chinese seemed not to have much knowledge. Florida varieties with multiple disease resistance were reported to do well in spring but not in summer. They suggest virus disease as the reason and resistant varieties just becoming available should certainly be tried. Varieties selected for fruit setting at high temperatures - such as 2029 from the Philippines, and Chico et al. from Weslaco, Texas - should be tried since the complete absence of tomatoes in the field so long before frost suggests a fruit set problem caused by high temperature.

2. Mosaic in squash (<u>C</u>. <u>pepo</u>). This disease, probably caused by cucumber mosaic virus is a serious problem. This has been confirmed by subsequent observation and by the fact that nearly all cucumbers introduced from China have been CMV resistant in the United States. No high level of resistance has been found in <u>C</u>. <u>pepo</u> but a modest level that may be useful exists in a plant introduction from Turkey used in breeding at Cornell University. This should be tested in China.

3. Much interest in cucumber breeding exists in China, but specific problems were not discussed. Talk centered on the use of gynoecious hybrids, with questions from Chinese scientists about how one maintains an all-female parent line suggesting that they have little experience with the character.

4. Questions on maintaining potato seed tubers without virus degeneration led to a long discussion of foundation and certified seed programs in the United States, which seem not to have counterparts in China. Mr. Yeh Fan had raised this question twice on the previous day of our visit to the Institute of Genetics. He stated that potato seed tubers can be produced in the Peking area by growing two crops per year - one planted in March or April and harvested the end of July, and one planted in August and harvested in November. Plants exhibiting virus disease symptoms are rogued from the spring crop, and tuber dormancy is broken for the August planting. Lower temperatures prevail during the growth of this crop and

are said to reduce the virulence of the viruses that may be present, making the tubers suitable for growing the next year. The recurrence of this subject suggests some doubt about the efficacy of the system. Similarly, the appearance of the potatoes in Kirin Province was not reassuring: the few we saw at close range appeared badly infected by virus and the fields we passed looked much less vigorous and green than one would expect at this time of year.

Vegetable Research in Shensi Province

The only vegetable research we saw in this province was at the Five-Star Brigade on the outskirts of Sian. Our time in the field was a few minutes in the rain as darkness was falling. Two vegetable specialists from the Shensi Academy of Agricultural Sciences at Wu-kung were stationed here to work with the peasant technicians and farmers in conducting research. We saw some good plots of inbred lines of Chinese cabbage and a test of experimental hybrid combinations. We also saw a cucumber planting related to breeding for powdery mildew resistance, but rain drove us from the field before we could see details.

Fortunately one of the vegetable specialists, Mr. Chao Shih-ya, accompanied us to the Northwest College of Agriculture the next day, giving us some time to talk. He said that in addition to the breeding we saw, they are working on the use of male sterility for producing hybrid cabbage seed. They are dealing with polygenic male sterility and best B lines give about 70 percent male sterile progenies. In their attempts to use self-incompatibility, their bud pollinations gave too few seeds.

Mr. Chao stated that in Shensi Province there are 320 production brigades (out of a total of about 25,000) that have experimental work on vegetables comparable to what we saw. These have 5000 peasant technicians, who are graduates of middle schools. This means they have about 12 years of school, finishing at age 18 or 19.

At the Northwest College of Agriculture, also at Wu-kung, we met with the members of the horticulture department, but were unable to see field work because of rain. Their main past achievement in vegetable breeding apparently was the development of the Shinung turnip, a very large variety

which sometimes reaches a weight of over six kilograms. It yields substantially more than standard varieties and has been widely used since
release in the late 1950's. It came from a cross between local varieties
from Shensi and Shantung Provinces.

Currently they are working on powdery mildew resistance in
cucumber and it was material from their program we had seen in the test
at the Five-Star Brigade. They collected cucumbers from all over China,
tested for mildew resistance, and now have some promising F_4 progenies.
Fusarium is also a problem with cucumbers and they have been searching
for resistance with some success.

They are also very much concerned about virus disease in tomatoes
and have been working with resistance from Holland since 1970. The level
of resistance has not been very good and we had a long discussion of the
higher level resistance at the TM-2 locus.

Perhaps it was because he had more time, he says, Dr. Munger
was especially impressed by the work being done on vegetables in Shensi
Province.

Vegetable Production and Research, Nanking, Kiangsu Province Agricultural
  Sciences Research Institute.

Wu-Kuang-Yuan is responsible for vegetables. He studied at
Cornell and Purdue in 1945-46, and works on tomato and onion. He is
assisted by Shu Hu-leng - tomatoes; Mrs. Li Wei-fung - Chinese cabbage; and
Ms. Yang Shu-ying - general vegetable work.

We did not visit any vegetable fields, either production or experimental, while in the Nanking area. Mr. Wu explained that their summer vegetable experiments had been harvested and fall vegetables were just being
planted.

In their research on tomatoes they are working on hybrids for
processing, especially of the San Marzano or paste type. Mr. Wu stated that
they have an experimental hybrid with alternaria leaf spot resistance which
he attributes to heterosis. They have Fusarium and Verticillium wilts but
apparently are not breeding for resistance. Growers are now using San
Marzano and #524, a selection from a native variety with large red fruit.
Nova, a new variety from Geneva, New York, should be tried here because it

is an early San Marzano type with fusarium and verticillium resistance.

They grow both red and yellow bulbing onions and are working on the use of male sterile inbreds to produce hybrids. Time did not permit learning the details of this work but it sounded as if it was in the early stages. We discussed the chive-like onion with flat leaves that we have seen at several locations. They identify it as <u>Allium</u> <u>tuberosum</u> and say it is mainly propagated by division in the fall but that seed can be planted to start a new crop in the spring.

In Chinese cabbage they are using self-incompatibility to produce hybrids that are just beginning to be used commercially. Again, limited time kept us from going into details. We did get clarification as to what is covered by the name Chinese cabbage. The term as translated includes both heading and nonheading types which have been classified as <u>Brassica</u> <u>pekinensis</u> and <u>Brassica</u> <u>chinensis</u>, respectively. They are now considered to belong to the same species. In the north, the main crop is the heading type, while around Nanking about two-thirds are nonheading. The latter can be grown every month and eaten at any stage while the former grows only in the fall. They also consider the nonheading type to be more nutritious because it is green. Data on composition confirm this.

Sweet potatoes are considered to be "food crop" and not a vegetable while the reverse is true of white potatoes. For the latter they are using the two-crop system to control degenerative virus diseases. The first crop is planted in early March and tubers for seed are harvested in early June, two weeks earlier than the commercial crop. The fall crop is planted in late August or early September and harvested in late November. Not much of this crop is grown for table use. Plants with virus evident are rogued from both crops. The success of this sytem varies with variety. Red Eyes, Good Harvest White, and White-Haired Old Man do not degenerate as much under this system as Nan Chou, causing one to wonder if some of their varieties carry more virus resistance than others.

Kiangsu Province (where Nanking is located) was reported to have a population of over 45 million in 1967, and was the most densely populated province. (Only a few countries in the world have a greater population than

- 133 -

this province.) As far as we can tell, most of the vegetables used by this population are produced within the province, and the Agricultural Research Institute is the main source of scientific information. The small group working on vegetables cannot possibly handle the amount of research that should be done for an industry and a population of this magnitude. A very large increase in the vegetable research in Kiangsu Province would seem to be thoroughly justified.

Shanghai Academy of Agricultural Sciences, Institute of Horticultural Research

The Horticultural Research Institute is one of five institutes of the Shanghai Academy of Agricultural Sciences.

The largest experimental planting we saw was cabbage. Here the objectives are to get greater cold resistance to extend the growing period into the winter and to improve uniformity, the latter mainly by selecting pure lines and producing hybrids between them. A number of institute workers are stationed in the communes and they, working with the peasants, select the most cold resistant plants. These are sent to the Institute for seed production which is done through natural intercrossing among the selected plants. Seed sown from individual plants is planted out for evaluation. Some of the resulting progenies were in the field and looked much more uniform than cabbage we saw in commercial fields.

Downy mildew is reported to be the most important cabbage disease in the area. Apparently no work is done on resistance.

There was one cucumber experiment in the field, and on inquiry we were told that they are breeding for downy mildew and Fusarium resistance. However, further inquiry brought the replies that no Fusarium infected plants could be seen and that downy mildew resistant lines were not in the planting; so its purpose was not clear. There was much downy mildew, some mosaic that was probably CMV, and a second and much more severe virus disease that was probably not CMV. All plants looked low in productivity with female flowers many nodes apart, but they said that all-female cucumbers have been tried and produce too many females and consequently too many cull fruits.

In the laboratory we were shown some data on vitamin C content of tomato varieties. They had been comparing hybrids with inbred parent lines. Both hybrids and parents had the same range from 26 to 37 mg/100 g.

The Aquila potato variety from Germany is the most important one in this area. Once again we find that the fall crop is used for seed as a means of virus control but here they say that virus diseases do constitute a serious problem. Asked about potato breeding in China, they said that some is being done at Hupei Agricultural University. They mentioned the K'e San Research Institute in Heilunkiang Province as the place from which they got Aquila in 1965 and the most likely place to be introducing further varieties from other countries. However, when a research institute in that province working on potatoes had been mentioned earlier in the trip, we were told it is no longer in existence.

They are starting to use F_1 tomatoes commercially in this area but not to any great extent as yet. Hybrids of other vegetables are not in commercial use.

No controlled inoculation is used in breeding for disease resistance.

No research is being done on the nonheading Chinese cabbage that is the main vegetable in all seasons here. They said this is because there are many satisfactory local varieties, yields are high, and there are no special problems. (Unfortunately we never had the chance to get into a field at or approaching harvest.)

General Comments on Vegetable Research

The amount of land devoted to vegetables in the areas visited, the quantity and nutritional value of vegetables in the Chinese diet, and the problems in vegetable production would justify a great deal more research on vegetables than we observed or heard about. There is a need to have much of this research on an interdisciplinary basis, involving physiologists, pathologists, entomologists, biochemists, etc. as well as horticulturists. Large numbers of peasant technicians can be effective in certain types of research but cannot do some of the essential and more basic research which requires teams of highly trained specialists.

Present research on vegetables suffers from lack of knowledge that is already available elsewhere in the world. Similarly, some of the best germ plasm is not utilized in breeding for lack of outside contacts and coordination within China. As a result work already done is being repeated or needed research neglected, and progress is consequently hindered.

- 135 -

A number of specific suggestions are incorporated in the recommendations on vegetables and fruits prepared at the request of our hosts and appended to this report. We understand that agricultural research itself is undergoing experimentation and change. The combination of needs in horticultural research plus the willingness to change could mean great progress in providing food for China's additional population in the years ahead.

Plant Physiology

We visited two institutes doing research in plant physiology: the Institute of Botany under the Academy of Sciences at Peking, and the Research Institute of Plant Physiology, Shanghai, formerly under the academy but now under the sponsorship of the Shanghai municipality. Several fields of interest were common to the two institutes; photosynthesis, nitrogen-fixation, plant hormones, tissue culture and microbiology. The work on tissue culture and microbiology is being presented in other sections of this report. Both visits were very brief and little detail can be given.

Photosynthesis

The photosynthesis laboratory of the Institute of Botany is conducting research in three separate areas: photophosphorylation, structure and function of reaction centers for photosynthesis, and the structure and function of chloroplasts. A method has been developed for the production of ATP from plant rather than animal sources. The method utilizes isolated spinach chloroplasts, ADP, an inorganic phosphorus source, and light. The method is being used commercially, and ATP is reported to be effective in the treatment of certain human liver ailments. Research has indicated two reaction sites in the chloroplasts involved in photosynthesis.

Single cell cultures from mesophyll tissue have been obtained following mechanical grinding. Such cells have been induced to divide and, in some cases, to produce callus tissue. Cytological examination of the cell masses indicates the chloroplasts have divided with at least some general correspondence between cell and chloroplast divisions.

At the Shanghai Institute work has centered on the resolution and reconstitution of chloroplasts with respect to "coupling" factors from different species. The photophosphorylation enzymes from different species are interchangeable but differ in reaction rates. The effects of fluorescence and high energy fields on photophosphorylation are also under study.

Plant Hormones

At the Institute of Botany three substances are under study: CCC, ethrel, and napthalene acetic acid. CCC is being used limitedly on tall varieties of rice to reduce plant height and thus contribute to lodging resistance. The effects of ethrel as a ripening agent have been explored with watermelons, peaches, and tomatoes at rates of 4000 ppm. Napthalene acetic acid has been used to some extent as a seed treatment for wheat. The treatment promotes sturdier seedlings and may be of value if wheat is being transplanted, an increasing practice in some cropping systems. An extract of water chestnuts has been effective in promoting the growth of pith tissue in cultures. The substance has not been identified and its possible effectiveness in the culture of other tissues remains to be established.

At the Shanghai Institute considerable research is devoted to an understanding of the factors involved in the shedding of squares and bolls of cotton. Gibberellins are effective in reducing shedding of squares when applied directly to flower buds. Foliar applications were much less effective and the cost prohibitive for field use.

Work reviewed at the Institute of Biochemistry indicated that nucleotides isolated from a wild yeast were effective in increasing yields of several crops. In tracer studies using P^{32} and K^{42} with rice, application of nucleotides at the rate of 40 ppm. stimulated root growth and increased the rate of nutrient absorption. Studies involving rape, cotton and potatoes indicated yield increases in the range of 5 to 10 percent.

Nitrogen Fixation

Both institutes are conducting research on nitrogenase. This enzyme has been isolated and separated into two compounds: one containing iron and the second containing molybdenum. Both of these components have been purified and crystallized. Neither component is active singly but activity is completely restored when the two components are mixed.

Some work is under way on the genetics of nitrogen fixing bacteria but neither the approach used nor results obtained were described. Attempts are also being made to infect nonleguminous plants with various strains of Rhizobia.

Phytotron

A new phytotron has been built at the Shanghai Institute.
Temperature and humidity controls are available and high intensity
lighting is provided by Xenon fixtures. These provide a light intensity
of 30,000 lux one meter from the light source and about 25,000 lux at a
two-meter distance.

The immediate use of this facility will be to study the effect
of temperature, humidity and day-length on flowering and seed development
in rice.

Department of Plant Physiology at the Academy of Sciences, Institute of Botany, Peking

Fruit and Vegetable Storage Laboratory.

Storage research with fruits and vegetables was begun in 1967
and has been done on apples, pears, tomatoes, and cucumbers. A storage
problem with pears was their tendency to rot at the core by February
when stored at $0^{\circ}C$ as was customary. It was found in research at this
laboratory that they keep well if the temperature is maintained at $5^{\circ}C$.
At the lower temperature the rot was caused by the activity of polyphenol
oxidase which gave rise to chlorogenic acid.

Controlled atmosphere storage has been found useful for tomatoes,
apples, and cucumbers. Details were given only for cucumbers, which they
said can be kept up to 45 days at room temperature if oxygen is lowered to
3 percent and CO_2 increased to 0.5 percent. This prevents seed development.
Farmers are using plastic to enclose stored cucumbers in China and in-
jecting nitrogen to lower the oxygen concentration.

It was stated that similar results have been obtained with tomatoes, contrary to most opinion in the United States. Unfortunately
time did not permit getting details of the tomato work.

Plant Pathology

Professional Society Affiliations for Plant Pathologists

Like other scientific societies related to agriculture, the
Chinese Phytopathological Society was disbanded after the Cultural Revolution. Scientists at agricultural colleges and the academies or
institutes for research, including plant pathologists, are now members of
an all-inclusive society, the Association of Agriculture of the People's
Republic of China.

One responsibility of the Association of Agriculture is to foster the advancement of the specific scientific disciplines in agriculture that are included in its membership. A directory of plant pathologists and their area of specialization is not available for China. In most colleges and institutes, plant pathologists and entomologists now work in departments of plant protection. Since many plant pathologists are engaged in broader programs in the general area of plant protection, it is difficult to designate certain individuals specifically as plant pathologists. In view of the total numbers of individuals assigned to disease control problems in the colleges, research institutes, county plant protection units, and communes, it is likely that more individuals are directly concerned with disease control problems than in any other nation in the world. One approximation was that about 50,000 persons are currently involved wholly or in part on plant pathology-related programs.

Publications and Dissemination of Research Findings

The Chinese Journal for Plant Pathology discontinued publication about 1966. Articles on plant virology and other areas in pathology are being published in Scientia Sinica and Acta Botanica (Sinica). In general, however, since the primary emphasis has been on applied aspects, most plant pathologists indicated that publications of their current work were either not available or work was still in preliminary stages. It was possible to obtain as gifts from the Association of Agriculture (Peking) a series of very fine extension-type manuals with color illustrations of the major diseases and insects affecting key crops. Although these were apparently designed for use at the communes, the illustrations usually include drawings of the reproductive stages of the causal fungi as well as details of the key stages in the life cycle of the major insect pests.

In addition, a detailed listing of the fungal pathogens affecting crops in the Kirin Province has been published recently, as was a listing of the chemicals that can be used for control of insects, weeds, and plant pathogens. This last manual describes the chemical structure and properties of these pesticides and lists the recommendations for proper use. Citations to these publications are listed in Appendix 6; these materials are in the library of the Department of Plant Pathology, University of Wisconsin, Madison.

The results of ongoing research in plant protection are disseminated in the following manner:

- Scientists, extension workers, and farmer groups visit the communes and brigades as well as various experiment stations at which a specific practice or advance in a control practice has been found to work effectively.

- Radio broadcasts are used extensively to disseminate specific information on controls.

- Meetings of workers in plant protection are held frequently and the relative merits of new practices are discussed and evaluated.

- A telephone network permits one call to alert a large number of communes to the prospect of a critical disease or insect problem.

Education of Plant Pathologists

The general program followed in the training of plant pathologists is similar to that for all students in agriculture. Students at agricultural colleges who will become plant protection staff members are enrolled in a Plant Protection Curriculum and are supposed to master both entomology and plant pathology at the basic level as well as spend one-third of their time in studies under field conditions. It should be emphasized that the specific programs of study are still undergoing evaluation and development. No new revised printed (published) teaching manuals or textbooks specifically for plant pathology are available as yet. However, each instructor may have some material available in duplicated form for the students, and texts previously in use are still serving as reference manuals. In one teaching laboratory, many large posters were displayed showing excellent color paintings of symptoms of disease and reproductive structures of important fungal pathogens. Similar posters were also available in the entomology classrooms. Relatively new, high quality binocular dissecting microscopes and standard microscopes were available for student use. Preserved specimens and large collections of carefully mounted and labeled diseased plant specimens and insects were available for student study.

Advanced training in plant pathology is limited to individuals selected by teachers and fellow students as worthy for further training; these individuals receive appointments as assistants and are trained by senior staff. The persons selected must be approved by the revolutionary committee of the college before being given an appointment as an assistant.

As indicated earlier, most agricultural colleges are in a transition period as they are now in the process of completing the moves from urban to rural areas. The Northwest College of Agriculture, initially established at Wu-kung (at some distance from Sian, the major city in the province) is one of the institutions that did not have to undergo this relocation process. Because this program of constructing new facilities and developing curricula is still in progress, it is not possible at this time to make any definitive statements as to the organizational or administrative pattern that will be followed for all plant pathology units in agricultural colleges. The likely model for most colleges is a department of plant protection with two divisions - one of plant pathology and one of entomology. Weed science is usually considered to be the responsibility of one or more members of the plant protection group but research on weed control may be under study in other departments as well.

Organization of a Basic Course in Plant Pathology

Selected individual diseases on the crops most important to the province are discussed in some detail. One important bacterial, fungal, and viral disease is studied in depth. Students make isolations and inoculations and complete Koch's postulates for at least one disease. Then students learn those aspects of mycology which will enable them to diagnose major diseases of economic crops. Emphasis is placed on diagnosis by symptomatology and examination of fruiting bodies of fungi. Basic aspects of epidemiology and control are discussed. In addition to the required courses in entomology and plant pathology, all students in the plant protection curriculum are required to take a course in chemical control that covers insecticides, fungicides, antibiotics, and weed control chemicals.

Virology

As is true for other specialized areas in plant pathology, most virology research is directed toward the development of improved methods of control. Demands on most professors of plant pathology have been so great that relatively few individuals have been in a position to specialize on more basic aspects of virology.

Virus diseases are relatively common but rarely appear to be the limiting factors in crop production except in a few instances. Intensive insect control efforts as well as a variety of cultural practices, including careful early removal of virus-infected plants and reduction of weed hosts in the vicinity of fields, may contribute to a general reduction of virus diseases. Careful selection of healthy plants for propagation or seed sources at each commune may enhance continued selection for resistant material as well as reduce hazard of dissemination of viruses: for instance, most native Chinese cucumber varieties are relatively resistant to mosaic.

In Kwangtung Province, virus diseases on sweet potatoes are relatively rare. Sweet potato mosaic was a problem in the province about 20 years ago but at present infected plants are rarely observed. Mosaic is common in cucumber and cabbage; a virus disease known as yellow top infects beans and peas, and turnip mosaic is often present. Bunchy top is a disease that is appearing on banana now. Papaya mosaic was rarely seen until very recently and now it is spreading widely in this province. Citrus yellow shoot is of growing concern and the exact nature of the complex that is described under this name has not been resolved. Mulberry stunt occurs but farmers do not consider that it causes enough losses to warrant efforts on control, and viral diseases of rice are fairly minor.

The Institute of Biochemistry in Shanghai conducts virology studies of rice black streak stunt, jujube witches broom, mulberry stunt, and citrus yellow shoot. Initial studies completed by Mr. Tien Shung-kung using electron microscopy indicate that mycoplasma-like organisms may be associated with the mulberry stunt disease and also with the citrus disease. Detailed studies on properties of the protein sub-units and nucleic acid components of citrus yellow shoot virus are in progress. This work has been published in Scientia Sinica 16:431-441.

Bacterial Plant Pathogens

Certain bacterial diseases are a source of real concern in China: the main problem is bacterial rice blight (Xanthomonas oryzae). Present efforts to control bacterial blight are directed toward the reduction of inoculum by the use of disease-free seed, avoidance of recontamination of seedlings, and development of disease resistance. Dr. Fang at the Kiangsu College of Agriculture in Yangchou has been the leading investigator of this as well as other related bacterial diseases such as bacterial streak (X. translucens f. sp. oryzicola). Through his studies on the importance of seed transmission of bacterial rice streak, strict quarantine measures were imposed that have resulted in the virtual elimination of this disease in Kwangtung and Kiangsu Provinces.

In Nanking, at the Kiangsu Academy of Agricultural Sciences, Madame Hsiao Ching-pu is using a bacteriophage specific for Xanthomonas oryzae in the following ways: detection of X. oryzae in seed; forecasting the increase of the pathogen in paddy water; and detection of the pathogen in crop residues.

Bacterial Wilt in Kwangtung Province

Bacterial wilt (Pseudomonas solanacearum) reportedly has occurred on peanuts, tomatoes, tobacco, and eggplant, but has not been observed on bananas. At present it is mainly a problem on tomatoes; because of bacterial wilt the farmers plant a late variety of tomatoes. It is difficult to explain why this has continued to be a problem on tomatoes when it no longer causes losses on peanuts. The system of alternating peanuts and rice has resulted in successful elimination of the disease as a major problem. However, if the rice field is not properly flooded, peanuts may be attacked. Studies on races and strains of P. solanacearum have not been made.

Bacterial wilt also occurs on tobacco in Honan Province but there the main disease problem is black shank.

General Status of Nematology

Relatively few persons are working solely on the biology and control of plant parasitic nematodes since these organisms are considered to be of minor economic importance in Chinese agriculture. Several nematode diseases were cited as the cause for some concern in the past. One of these was the wheat gall nematode which is now under control. The soybean cyst nematode does cause a problem in localized areas and peanuts are affected by root knot in Shantung Province.

At the Kiangsu Academy of Agricultural Sciences (Nanking) Mrs. Shih Wen-ying is studying the white tip disease of rice caused by a species of Aphelenchoides oryzae. Apparently this nematode was introduced into China on seed from Japan in 1933. The spread of the nematode is restricted by the imposition of strict quarantines. This nematode was present in Kwangtung Province in the 1950's but it has apparently been eliminated by stringent quarantine restrictions and cultural practices.

At the Kwangtung College of Agriculture projects are now being developed to assess the impact of nematodes on several major crops. Dr. H. C. Faan (head of plant pathology) commented that he had observed root knot on greenhouse grown tobacco seedlings soon after his return to China in 1950, but that he had not observed it as a field problem on tobacco at any time. He did mention some nematode damage to mulberry and flax.

Recognition of certain factors may minimize the effect of plant parasitic nematodes on crops in China: many of the soil types are not favorable for nematode development; the alternation of rice paddies with other crops results in unfavorable growth conditions; the use of organic fertilizer probably favors the development of antagonistic organisms; and shifting many different crops in sequence, as in the case of certain vegetable crop sequences, may be unfavorable to nematode development.

Disease Forecasting and Protection Units

One of the most important components of the system to reduce losses from insects and diseases is the County Plant Protection Unit. These units were established in the post-liberation period. Each county has a team of 4 to 8 technically trained persons who are engaged in surveys of insects and diseases. They maintain observation plots at the county experimental field stations and keep a close watch on the development of insects and diseases. If it is apparent that a problem is developing that can be controlled by timely applications of a spray or dust, the plant protection specialist can make a conference call through a telephone network to all farm communes in the county and also alert neighboring counties. Some of the individuals employed in county units have college degrees in plant protection. Although most of the units' work is directed to insect control, these groups also provide very effective surveillance on the outbreaks of diseases.

If new or unusual pests or diseases appear, the county workers can and do call for assistance from the provincial agricultural academy or college. Once appropriate recommendations are determined, they can make immediate contacts with communes and also make announcements on the radio. In addition to survey responsibilities, some protection units may evaluate pesticides on an experimental basis and assist in dissemination of new information on disease and insect control.

Each county has an individual who provides certification slips for shipment of seed out of the county. Reportedly the strict application of a quarantine involving county personnel made it possible to eliminate bacterial rice streak from Kwangtung Province. No commune on which the disease appeared was permitted to sell its seed and the only seed permitted to move from one county to the next was that certified as free of bacterial streak. Since it had been established that bacterial streak was seed-borne and did not survive readily in the soil between crops, the application of strict quarantine methods proved to be highly effective for this seed-borne pathogen.

Although it has not been possible to use this same procedure as effectively in the control of bacterial rice blight caused by Xanthomonas oryzae, use of inspection procedures and disease-free seed is an important component of control for this disease also.

Ecology of Soil-Borne Pathogens

In view of the intensive cropping practices that are followed in most regions of China, it would seem likely that root-rot diseases or soil-borne pathogens would be a major limiting factor in crop production.

In fact root rots appear to be a relatively minor concern. Damping-off caused by either Pythium or Rhizoctonia may occur, but losses are considered to be minor. In the vicinity of Canton, Sclerotium rolfsii was observed causing some damage on eggplant, but apparently actual loss from this pathogen also is relatively little. Fusarium wilt of cotton is a major problem and pathologists in 17 provinces have a Fusarium wilt working group established that is amassing all available information in an effort to resolve the problem.

Most of the studies on control of root disease problems have been directed at modification of cultural practices rather than analysis of the population dynamics of the pathogens involved. In those areas where specific root-rot problems have occurred, peasant farmers have either shifted away from the susceptible crop or have changed planting times.

It would appear that the practice of alternate flooding and drying as well as the type of organic fertilizer used are key factors in minimizing losses. In the case of vegetable crops the succession of crops is so varied and the

growth period so short that root diseases may have little opportunity to inten-
sify. It is also possible that because of centuries of intensive cultivation,
many of the local varieties have developed a high tolerance to specific root
disease pathogens.

It would be of interest to conduct a detailed study of the population
dynamics of a few selected root pathogens under the cropping system that has
evolved in south China.

Basic research on disease physiology, mechanisms of pathogenesis and
the biochemical nature of disease resistance is not in progress at any of the
institutes or colleges that we visited. We were informed at the Institute of
Biochemistry in Shanghai that they were not aware of studies on the biochemical
basis of resistance. Due to limited time, it was not possible to visit the Institute of Microbiology in Peking which has been combined with the Institute of
Applied Mycology.

Chemical Control of Plant Pathogens

Organic mercury compounds were once commonly used as seed treatment
materials. These have been placed under restriction recently and once current
supplies are used, no further manufacture of these compounds is planned.
Oxithiin is used as a seed treatment for control of smut diseases of wheat, and
seed treatment with organic mercury has been recommended for control of the
Bakanae disease caused by Gibberella fujikuori. A few fields were seen in which
the Bakanae disease was present; these were in Kwangtung Province. Currently,
formaldehyde is recommended and the disease can be controlled effectively when
the formalin is properly applied.

Topsin-M (#157) is apparently being used for control of a wide range
of leaf spot diseases. Bordeaux mixture is still commonly used for control of
many foliage diseases. However, zineb and maneb are also used on some crops.

In Kirin, two antibiotics which can be manufactured locally by use of
the appropriate Streptomyces culture have been tested for control of rice blast
(Hinozan and Kasugamycin).

Antibiotics for Disease Control

At the Institute of Plant Physiology (Shanghai) the Laboratory of
Microbiology has a study in progress on the development of an antibiotic that
shows promise for control of rice blast. Using a strain of Streptomyces aureus
a product named Gougeratin can be obtained which is recommended for two applications to obtain effective control. Procedures for production of the antibiotic

have been simplified so that the medium can be obtained and the antibiotic produced at the commune level. Industrial production of the compound is under way.
The structure of the product has not been determined but it was described as an
alkaline soluble nucleotide.

Insecticides

Use of insecticides is still the common and is the sole means of
controlling some pests and virus vectors. The use of DDT and other chlorinated
hydrocarbons has been placed under legal restrictions. An insecticide in common
use is 666 (benzene hexachloride). Apparently there is deep concern about the
continued application of this chemical. Once current stocks of many of these
insecticides are used, no further supplies will be provided.

The three major insecticides in use in Kirin as well as many other
provinces were:

Phoxina - for control of aphids, May beetles, and rice borers

Dipterex - for control of army worms

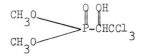

Rogor - for control of aphids

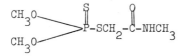

Use of 666 for general insect control is still widespread (benzene
hexachloride - $C_6H_6Cl_6$.)

Cultural Practices

Seed certification. With the emphasis on self-sufficiency of individual communes, the practice of producing seed at each commune has had certain
beneficial aspects. For one thing it has reduced the possibility of disease
spread by seed from a common seed source in which care to insure quality of
seed was not practiced. However, in the case of certain virus diseases, it is
unlikely that the level of sophistication requisite for detection of viruses
is present to insure that virus-free material is retained for seed or for
propagation.

The practice followed by farmers for maintaining "virus-free" potato
tubers for seed is an interesting and puzzling phenomenon. Apparently potatoes
from the spring planting are harvested slightly prior to the main harvest and

tubers are retained only from carefully selected disease-free plants. These
tubers are then used to establish a fall crop which is also harvested slightly
before the normal harvest period. Tubers from this fall harvest serve to
plant the next year's crop. It is difficult to understand why this method has
been reasonably successful in the elimination of viruses (if it has). The
fact that there was concern about losses from ring rot indicates that the seed
selection method has not worked effectively for this disease.

Breeding for Disease Resistance

The major emphasis in each of the departments visited was on the
search for increased levels of resistance to the major diseases. This work was
done in collaboration with the respective plant breeders. Since each of the
communes has a group concerned with selection and crossing and developing improved varieties, the total effort within each region on the specific major
crops is enormous. Relatively little work is in progress that involves the
use of artificial inoculation under controlled conditions in greenhouses or
controlled environmental conditions. Some field inoculations were being conducted at Kung-chu-ling with H. turcicum on corn and also with Fusarium
oxysporum f. vasinfectum on cotton at the Northwest College of Agriculture.

As each commune includes a group concerned with seed improvement, it
is possible for the experiment stations in each province to test lines with
improved disease resistance over a complete range of environmental conditions
and also at locations where fairly intense disease conditions prevail. Seed
of new improved varieties can also be increased rapidly and spread over millions of acres in a given province in a very short period of time.

In Kirin Province the severity of rice blast and the appearance of
new races has been reduced because varieties with four different sources of
resistance are being planted under the supervision of the academy at Kung-chu-
ling. With the control system at the county level and supervisory staff available, this practice can be implemented very effectively. Similar procedures
are followed in the planting of poplar clones. Apparently four or more different clones are being used in the propagation of cuttings.

Prevalent Diseases, By Crop

    Millet

Xanthomonas translucens	bacterial streak
Sclerospora graminicola	downy mildew
Ustilago neglecta	smut
Uromyces setariae	rust
Piricularia grisea	blast
Corticium sasakii	sheath blight
Red leaf virus	

Of the millet diseases, blast is the most damaging, downy mildew is second, and rust (Uromyces) ranks third.

    Corn

Helminthosporium turcicum	Northern leaf blight
Helminthosporium maydis	Southern leaf blight
Fusarium graminicola	stalk rot
Ustilago maydis	corn smut
Spacelotheca reiliana	head smut
Kabatiella zeae	frogeye leaf spot

The fungus causing the most damage was H. turcicum; certain lines were very severely affected. The second most important disease was caused by Spacelotheca reiliana, a pathogen rarely seen on corn in the United States. H. turcicum had only recently become a problem; the increased incidence of the disease was associated with the use of certain inbred lines that had not previously been used. Their most resistant material showed evidence of a fair level of infection and it is likely that U.S. lines with higher levels of resistance could be used to some advantage in China. Frogeye leaf spot was not observed as a problem in Kirin until the late 1960's; nematode problems on corn, or on the other major crops, were not a matter of concern.

    Soybean

Alternaria tenuis	Alternaria leaf spot
Cercospora sojina	frogeye leaf spot
Cercospora kikuchii	purple stain
Mycosphaerella sojae	Mycosphaerella leaf spot
Peronospora manshurica	downy mildew
Phyllosticta sojaecola	Phyllosticta leaf spot

Xanthomonas phaseoli var. sojae	bacterial pustule
Ascochyta glycines	Ascochyta blight
Septoria glycines	brown spot
Sclerotinia sclerotiorum	stem rot
Pseudomonas glycines	bacterial blight
Soybean mosaic	

Mosaic, and downy mildew, and two bacterial diseases, are considered to be the most destructive of the diseases at present. Nematodes are not cause for concern except for localized outbreaks of the soybean cyst nematode in certain western areas.

Sorghum

We saw relatively few diseases of sorghum. Most prevalent was sooty stripe (Ramulispora sorghi). This produces a large leaf spot similar to H. turcicum in size and appearance. A second leaf spot, Cercospora sorghi was also present. Kabatiella zeae has also appeared recently as an important leaf spot on sorghum.

Castor Bean

Infection by Pseudomonas ricini was present in a planting of castor bean, but apparently it was not causing much damage. Castor beans are planted along the edges of fields in most of the areas of north, northwest, and mid-central China.

Wheat (See section on wheat, pages 54-64)

Observations on Plant Diseases, Nanking, Kiangsu Province, Hung-ch'i
  (Red Flag) Commune

        Downy mildew was very severe on cabbage. TMV and CMV were
present on tomatoes. Each of the tomato flowers in this field was being
treated by hand with a solution containing a low concentration of 2,4-D
to prevent flower abcission. This obviously involves a great deal of labor
as the plants are continually flowering. A red dye is incorporated in the
2,4-D solution so that treated flowers are identified. Tomatoes are sprayed
with Bordeaux mixture to prevent Alternaria leaf spot. Ring rot was one
of their serious problems on potatoes. The commune grows its own seed
tubers and follows the practice of using fall grown potatoes as the source
of seed for the next season, as is true in all other areas. Late blight
was controlled by the use of a resistant variety and Bordeaux mixture.

        The chemical mainly used for insect control on the commune was
Dipterex.

Academy of Agricultural Science, Shanghai

                Key Pathogens
Mandarin oranges   Elsinoe fawcettii
Peaches   Gloeosporium laeticolor
Pear   Physalospora piricola, Venturia pyrina. Bordeaux
                      mixture was used for control of the pear diseases.
Cabbage   Downy mildew was a most severe problem. Zinc dithane
                      was used for control. Xanthomonas campestris was
                      not a problem.

        They mentioned a long history of study of the potato at this
institute. In 1965 they obtained the Aquila variety and this has been used
as a base for selection and improvement. Presently ring rot and viruses are
receiving attention. As is true in other areas they use disease-free tubers
from fall grown crops as their seed source. They consider that Aquila gives
them some protection against late blight.

Cotton Diseases (See section on cotton, page 108.)

        The main concern in the major cotton growing provinces is
Fusarium wilt. A coordinated program has been established involving approximately 30 scientists in 17 provinces. The program is coordinated by the

Northwest College of Agriculture and Forestry (Sian). They have obtained a large collection of isolates from the entire region and are subjecting breeding lines to inoculation in the seedling stage and also under field conditions.

In the Nanking area, Mr. Kuo Tsung-ch'ien, the cotton specialist at the Kiangsu Academy of Agricultural and Forestry Sciences, indicated that boll rots were a major problem. Of these, the most important was the Phytophthora boll rot. They were also concerned with obtaining resistance to Fusarium wilt and were cooperating in the project with the academy at Wu-kung, Shensi Province.

Use of Fungi as Food and Source of Medicinal Products

Mushrooms are an important component of the Chinese diet. At least one type of mushroom is served at most meals.

In the Shanghai area many communes are engaged in large scale production of Agaricus bisporus f. albida. A certain percentage of the local mushroom production is apparently shipped abroad. This production has increased greatly in recent years.

One commune close to Shanghai that we visited reported that they had produced 375 tons of mushrooms (mainly Agaricus bisporus f. albida) in 1973. Since there are about 190 communes in the Shanghai Province, many of which may also be involved in mushroom production, total tonnage for this region alone must be quite impressive.

Studies were in progress on the following fungi which are grown in the area for food: Agaricus bisporus f. albida; Lentinus shiitake; Volvaria violacea; and Auricularia judae. All were under cultivation in the communes and we were told that a significant portion goes to foreign markets. In addition they were growing Tremella fusciformis and Ganoderma lucidum, and use them as drug sources. Ganoderma lucidum is grown mainly in culture, whereas Tremella fusciformis is grown on logs of Pterocarya stenoptera. Ganoderma lucidum extracts are used as a means of reducing high blood pressure. Lentinus shiitake is grown in logs and culture. Apparently they had relatively few problems in growing Agaricus bisporus until very recently. Current problems may be attributable to viruses, but the specific agents or viruses involved have not been identified. An insect that affects the white

Tremella fusciformis is a cause of increasing concern. Mushroom growing is a large industry in itself and provides a valuable supplement for commune incomes in this region. In our drive through the area we noted many mushroom houses with the typical elongated chimney vents.

Zizania plants infected by the smut fungus, Ustilago zizaniae, are propagated as a special food crop in Kiangsu and Kwangtung Provinces. The plants have an enlarged fleshy swollen base in the area invaded by the mycelium of the parasite and infected plants are harvested before spore production is initiated.

Weed Science and Weed Killers

Under the constraints of our schedule, it was difficult to make an assessment of the current status of research on weed control.

The availability of manpower, normal cultural practices, alternate flooding in paddy rice culture, and cropping sequences all lead to a reduction in weed populations. The borders of fields and the roadsides are also constantly cut or grazed by farm animals and this reduces seed production. Most fields were remarkably free of weeds and there was evidence of frequent weeding by hand.

Weed killers are used limitedly for certain weed pests. Since most of the plant protection units we visited did not include weed control specialists, it can be assumed that this area of research has a low priority. The course in chemical control of pests at one agricultural college does include a treatment of weed control chemicals, and in the laboratory for weed control research, Institute of Botany of the the Academia Sinica (Peking), research is conducted on chemical weed control.

Common U.S. herbicides have been evaluated on rice, corn, and several vegetable crops. MCPA and DCPA (2,4-D) have been found to be effective alone or in combination as preemergence treatments for broadleaved weeds in paddy rice. We were told in this laboratory that the use of 2,4-D is fairly widespread in some communes, but the extent to which this is practiced was not determined. Other chemicals being tested as herbicides included Dalapon and related materials currently in use in the United States.

Potential for Germ Plasm Exchange

The team found Chinese scientists and administrators quite
interested in establishing a germ plasm exchange program. From their
viewpoint, it fits in well with their production-oriented research. From
our standpoint, it is equally desirable in terms of basic genetic resources
of China which are largely lacking in our base of variability. There is a
degree of urgency in developing these potentials in view of the speed with
which the Chinese are shifting from their conventional peasant selections
to hybrids. In discussing this subject we have divided it into two sections:
China as a center of origin; and the mechanics for plant germ plasm exchange.
At the outset we point out that the exchange is already underway as the
team brought a number of seed lines and a few plant materials to China and
in return received seeds and a few plants at each location visited (see
Appendix 5). Moreover, a team of Chinese agricultural scientists visited
the United States in September 1974, bringing seed samples and taking some
U.S. materials back to China.

China as a Center of Origin

Agricultural civilization in China has been traced to the loess
deposits north of the Tsinling Mountain range: the Yang-shao ("painted
pottery people"), discovered by J. G. Andersson, a Swedish archaeologist,
in 1921. The best known and most complete excavation is the site of Pan-p'o
which the Plant Studies Delegation was able to see while at Sian. An excellent museum including a roofed-over segment of the site has been erected
and tells the story of China's first farmers. Though radio-carbon dates
place the village at about 4000 B.C., there are indications that this civilization may date considerably earlier.

*　　*　　*

Setaria and Panicum millets were the staples of the early north Chinese agriculture. These were supplemented by gathering wild plant food, hunting, and pig rearing. Seeds of Chinese cabbage found at Pan-p'o foreshadowed the importance of Brassica in the modern Chinese diet. The millets remained central to Chinese culture into historical times. The legendary ancestor of the western Chou tribe was the god of millets. An early Chinese unit of measure was the length of 10 millet seeds, and units of weight were based on the weight of millet seeds. Of the five grains traditionally sown by the emperor in a yearly ceremonial planting rite, three of them were originally millets, one was rice, and the other wheat. Millet was the foundation of the Yang-shao culture and remained the basic crop in north China until recent times.

Archaeologically, a series of cultures has been identified leading from Yang-shao through several intermediate stages to Lungshanoid and then to the fully developed Lungshan Neolithic culture (the "black pottery people" of Andersson). These cultures diffused from the loess highlands of Shensi, Shansi, and Honan onto the eastern coastal plain and southward to Fukien. One Lungshan site was excavated on Taiwan by K. C. Chang of Yale and radiocarbon dated to about 2300 B.C. In Lungshan times, pottery was wheel-turned, cattle were domesticated and rice was a staple crop. Remains of rice have been identified in Lungshanoid sites dating to about 4000 B.C., but these sites are near the Yangtze and south of the Tsinling Mountains. Rice has been found at Non Nok Tha in Thailand at about the same time or possibly earlier, so the place or places of rice domestication in Asia have not been identified with any certainty. In general, Neolithic cultures south of the Tsinling Mountains are not well documented and connections, if any, between north and south China at that time have not been established.

It should be noted that the Chinese Neolithic, as now understood, is in some aspects late compared to the Near East, the Americas, and possibly Southeast Asia. In Palestine, Turkey, Iraq, and Iran, well established farming village communities date to 7000 B.C. or earlier. Evidence of plant domestication in Mexico can be traced to at least 6000 B.C. and fully domesticated beans and lima beans have been found in Peru also dating to about 6000 B.C. Studies of the Neolithic period in Southeast Asia are just beginning but rice culture as early as 6000 B.C. has been claimed on indirect evidence. The relative age of Chinese agriculture is less important, however, than the fact that it developed independently of other centers and produced its own unique set of plant domesticates.

China was listed as one of eight centers of origin by N.I. Valilov
who attributed over one hundred cultigens to the region. Some of these are
very minor and many were actually domesticated elsewhere but developed secondary
centers of variation under the skilled guidance of Chinese peasant farmers.
Among domesticates that appear to be Chinese in origin are a complex of Brassica
species, representing various kinds of cabbages, kales, and the like, some
endemic Allium (onion) species, the soybeans, Sagittaria, the smut-infested
Zizania, endemic races of cucurbits like Benicassa, Luffa, Trichosanthes and
Cucumis, as well as radish, Ipomea, Amaranthus, and Malva.

Persimmon, peach, jujube, several forms of Citrus, and special races of
chestnut, hazelnut and walnut were domesticated locally. Ramie, tung, lac,
bamboos and silkworms raised on mulberry leaves are technical crops of importance,
while tea and several cultured fungi are also Chinese domesticates. Gardens and
parks of the world have been enriched by Chinese contributions such as the Ginkgo,
daylily, chrysanthemums, and selections of camellias, azaleas, rhododendrons,
flowering crabapple, flowering pear, flowering peach and many others. Valuable
turfgrasses include Zoysia and centipede grass.

China as a Center of Diversity

One of the most conspicuous features of Chinese agriculture, however,
is the facility with which new crops are accepted and integrated into the system.
Chinese agriculture, as we see it today, relies heavily on introductions from all
over the world. Wheat and barley reached China during the Shan dynasty when the
capital was at Anyang (before 1027 B.C.). Over a period of succeeding centuries
other Near Eastern crops found their way to China, some are documented in the
literature, others are not. Introductions from the west were accelerated after
the Greek states were established in Asia following Alexander's march to India.
These served as termini for the overland silk routes, trading between China and
the Mediterranean world.

Imports from the west included wheat, barley, oats, flax, pea, horsebean, lentil, vetch, carrot, beet, lettuce, cabbage, cauliflower, parsley, celery,
alfalfa, and probably rape, although independent domestication is also possible
in this case. Radish and turnip are probably confounded by having both imported
and endemic races.

From India came sesame, eggplant, sugarcane, and some races of cucumber.
The luffas and bitter gourds may also have come by this route. The African
domesticates are few but important and include sorghum, cowpea, and watermelon.

Lablab niger is also wild in Africa but the Chinese races may have come from south Asian sources.

The greatest exotic contributions, however, were domesticated by Indians of the Americas. These include maize, sweet potato, potato, common bean, upland cotton, sea island cotton, peanut, tomato, peppers (hot and sweet), Cucurbita moschata, C. maxima, C. pepo, lima bean, papaya, pineapple, sunflower, and probably others we have not seen.

Although there are reports in the literature of peanuts found in Chinese Neolithic sites, the arrival of most of the American crops (including the peanut) is documented in Chinese literature and dates, generally, to the 16th century A.D.

Genetic resources in China, therefore, are a reflection of the history of Chinese agriculture. According to present evidence, we may visualize a nuclear area on the loess north of the Tsinling Mountains. There, a millet-based agriculture developed, flourished, and spread taking up new crops as time went on. Temperate crops were developed from the local flora such as Malva (according to Li, the most important vegetable of ancient China), Brassica, Amaranthus, Benincasa, turnip, radish, Chinese onions, persimmon, jujube, peach, and so on. Soybeans were added from the northeast (perhaps) and rice from the south. Water plants from the swampy coastal plain were exploited such as Lotus, Zizania, Sagittaria, Eleocharis, Tropa, water convolvulus and a number of others. This complex was joined by other contributions from the south: bitter gourd, luffa, cucumber, taro, Dioscora esculenta, D. alota, Lablab, Canavalia, ginger, litchi, tung oil, ramie, tea, Citrus and more, all rich in diversity either in south China or in adjacent regions.

Then, the plants from the Near East arrived and wheat and barley became staples in the cool northwest. According to Vavilov, many of the Near Eastern crops developed secondary centers of diversity in China. He particularly mentioned wheat, barley, oat, flax, pea, horsebean and lettuce. The group as a whole has been isolated from its original homeland for as much as three millennia, and this is sufficient for the development of unique landrace populations. Sorghum probably arrived in China somewhat later (see section on sorghum), but the Chinese kaoliangs are distinct in some respects from all other races of the crop.

The American crops have been in China only about 400 years and it is unlikely that the introductions included very broad genetic bases. Although they are now immensely important to Chinese agriculture and contribute a great deal of food and fiber for the Chinese, they are less interesting as genetic resources. Nevertheless, 400 or more generations in annual crops are adequate to establish new and special gene combinations and frequencies that could be of value.

The waxy corn gene was first identified in Chinese material and it is quite possible that new genes for resistance to some diseases have appeared.

Altogether, China is (or was) enormously rich in genetic resources, most especially for ancient indigenous crops, native flora, and the older importations from western and southern Asia. This is the primary reason for the interest of the U.S. Plant Studies Delegation in germ plasm preservation and exchange.

Genetic Erosion in China

It is impossible to assess in any detail the status of genetic erosion in a country as vast as the People's Republic of China in a four-week study tour. It is all too obvious, however, that erosion is far advanced. The new, integrated, three-in-one programs are functioning. A technology sufficient for the development of hybrid corn and sorghum has been instituted at the production brigade and production team levels. Hybrids have largely replaced the open-pollinated or landrace populations of both crops. Furthermore, corn is replacing sorghum in the northeast and is sharply reducing both sorghum and millet on the irrigated lands of the northwest. On dryland both sorghum and millet have an advantage and will persist, but in Shensi we already had to search to find the old types of kaoliangs. New hybrids had replaced all but a few remnants of the old ones.

Old landrace populations of wheat are being replaced by short-season, short-strawed, fertilizer responsive varieties. The emphasis on more crops per year, is a strong incentive to discard the old types and to install shorter season, earlier maturing ones. Rice follows an identical pattern. We did not see any of the barleys but the same route is inevitable for this crop.

The entire agricultural milieu is permeated with "hurry up," get one crop out of the way as fast as possible in order to get the next one in. Barley may be interplanted with cotton a month before harvest in order to get cotton started early enough that it can be picked in time to follow with rice (Nanking-Shanghai area). There is extreme and constant pressure for shorter season, earlier maturing varieties that are short stalked, lodging resistant, fertilizer responsive, adapted to close planting, and so on. The pressure can only increase with the installation of the projected new large capacity fertilizer plants. The

- 158 -

old populations of most crops, both indigenous and introduced, have already been largely replaced by new products of plant breeding.

All of this is understandable and commendable, but unfortunately little or no attention has been given to the conservation of traditional germ plasm. Indeed, at times enthusiasm for the new has perhaps overshot any rational process of transformation. There is no reason, for example, to replace some of the fine old apple varieties with "Golden Delicious" or "Starking," but China is not alone in this respect. One day we may wake up to find we have only two or three varieties of apple in commercial production in the whole world.

Attempts at preserving germ plasm appear to be few and sporadic. We did learn of two sizable collections of rice, one for japonica and one for indica types, but were unable to detect any national policy, planned program or even concern for the preservation of genetic resources.

The emphasis on local self-reliance recommended by Chairman Mao in "the four selfs" (self selection, self production, self use, self storage) related to seed at the production brigade level could have serious repercussions in the long run. The policy tends to build very narrow genetic bases and increases genetic vulnerability. In general, Chinese plant breeders have not viewed germ plasm from a broad, global perspective and have only worked with an extremely limited range of the total genetic resources available for any given crop. This tendency is accentuated when breeding programs are set up at the brigade or commune level.

There are, however, a few features tending in the opposite direction: If each brigade tends to produce somewhat different hybrids or selects somewhat different lines, the total genetic diversity in a commune, a county or a province would be maintained at a higher level than if a few hybrids or selections were used over a wider area; some communes have a policy that at least four cultivars of each major crop should be grown on the commune, thereby maintaining some diversity and somewhat reducing chances of epiphytotics; the intricate inter-planting patterns involving different crops or even alternate strips of tall and short cultivars of the same crop should also reduce the likelihood of epiphytotics; close identification with the masses in research work may tend to restrain extremely radical deviations from proven materials and methods as peasants tend to be cautious and prudent by nature; multiple uses of a crop, e.g. soybeans for dry beans, green beans and fodder or peas for dry peas, green peas, edible pods and pea sprouts, and so on, result in more varieties and a broader genetic base than for a single-use crop.

These reverse or stabilizing trends may have some value in maintaining genetic diversity and reducing genetic vulnerability, but we must conclude that genetic erosion is far advanced in the People's Republic of China and the trend will continue for some time.

Mechanics for Germ Plasm Exchange

Throughout our trip we noticed the marked shift to advanced lines. The encouragement to select improved seeds given by Chairman Mao is vigorously pursued; as a consequence, the early landraces are being replaced. However, there is sufficient individual commune and brigade preference as well as somewhat conservative attitudes that still allow for diversity from region to region. The greatest threat will be to wild types and varieties with long growing seasons that do not fit into the newer patterns of multiple cropping. Exactly how this will shift the nature of germ plasm diversity is difficult to predict but it will certainly narrow the base. It also seems that Chinese agricultural administrators have not given this matter much consideration. Apparently there was an attempt in 1966 to collect landraces at the communes and this may account for the figure we have seen that suggested a collection of some 200,000 samples of 53 crops. There is no question about the number of crops grown and the figure is low since at one vegetable commune alone over 60 types of vegetables were said to be cultivated. As for the germ plasm base, it does not exist as a solid unit. Rather, there is some degree of local collecting and conserving of seed at the brigade level. This is sporadic and it was reported that duplication under different names as well as several varieties in a single collection lot encouraged abandonment of a large-scale effort.

In meeting with Professor Wu Chung-lun and Ms. Tung Y. C. (Institute of Agriculture, Peking), we learned that China would indeed wish to exchange germ plasm with American scientists. Although there is no formal Chinese plant introduction unit at present, the Association of Agriculture (Peking), under the leadership of Mr. Hao Chung-shi, Acting Vice-President, serves as the best national contact for germ plasm exchange.

Scientists may wish to write directly to the institutions visited by the team and these are listed in the appendix. It would be advisable to send a copy of such correspondence to the Association for Agriculture, Peking. In addition, scientists will need to follow U.S. Plant Quarantine Regulations with respect to materials requested from China. To be advised on proper quarantine-introduction procedures, the matter should be brought to the attention of

- 160 -

Mr. Howard Hyland, Plant Introduction Officer, Plant Genetics and Germ Plasm
Institute Agricultural Research Center, Beltsville, Maryland 20705.

As a result of the visit of the U.S. Plant Science Group, the Chinese
institutes visited are prepared to engage in seed and plant exchange and persons
seeking to initiate a contact in China would do well to refer to the visit of our
group.

Plant Exploration

The prospects for Americans to engage in plant collecting in China are
not good for the near future. Problems of travel, somewhat limited accommodations,
and shortages of interpreters necessary for extended travels off the regular
route would need to be resolved before American scientists could expect to receive
individual invitations. The Chinese are not very active in germ plasm collection,
although there is some collecting of wild fruits in the mountains, and wild species
with herb potentials.

SOCIAL AND POLITICAL FACTORS AFFECTING AGRICULTURE

Since Liberation, rural China has experienced a revolution in local administration. Essentially this consists of the extension of the network of national bureaucratic control down to the sub-county level, and the amalgamation of political and economic management at that level. The new primary sub-county unit is the People's Commune (PC), which has consolidated the lowest level of bureaucratic management at the scale of the old marketing community. Under the Empire and Nationalist regimes, the county was the lowest level manned by professional administrators. Barely adequate to collect taxes and maintain order in a static economy and traditional polity, the county unit proved unable to govern an expanding population increasingly affected by modern economic and political forces. The commune, however, is a unit that is small enough to be able to amalgamate economic and political management and, more important, to coordinate the work of a purposively developing society. Commune administrators are national cadres (i.e., on the state payroll) and although they seem to be largely drawn from the local area, they can be transferred as the need arises. Thus their basic accountability is to the national bureaucracy, though they are close enough to their local areas to be able to understand their problems and communicate effectively with their people. The impact of this layer of officials for agricultural development is obviously crucial. Those that we observed - primarily commune administrative managers (pan-shih-tsu fu-tse-jen) - were impressive in their detailed grasp of the economic and social realities of their units. If the quality of these administrators is generally as high as what we observed, they are a major national asset in China's effort to modernize agriculture.

The next lower level, the one with which these PC administrators must deal, is the production brigade. The leaders of these units are literate peasants who are directly involved in production but are not on the state payroll. Each brigade leader has administrative responsibility for a number of production teams that are essentially congruent with the old natural village units. The brigade coordinates some important basic services such as the local primary school, the medical dispensary, and basic-level political work. The team influences the individual's life most directly, since it coordinates labor, and in most cases is the accounting unit that computes work-points and grain-rations, thereby governing peasant livelihood. However, the basic services at the brigade level mean that the peasant has a wider community to relate to on a daily basis. Peasants and team leaders seldom have occasion to deal directly with commune headquarters, though brigade leaders do so regularly.

This network of communication and accountability is a potent instrument for agricultural development. Not only can it allocate resources on a planned basis, it can also serve as the transmission belt for new agricultural technology. China's new rural administrative structure - responsive to local conditions as well as national programs - forms the basis for the drive to modernize agriculture. The displacement of the old elite controlling group and their village agents has made possible the present administrative system on the sub-county level. Thus the political revolution itself was a prerequisite for economic development.

Administrative Aspects of Agricultural Extension

Since the early years of the 20th century, China's effort to modernize agriculture and village society has made use of the "model" or "experimental" unit to propagate advanced techniques. The theory was that success in one area would evoke emulation by others. There were a number of "model county" programs during the 1920's and 1930's, none of which was able to spread its influence because the basic socio-political problem remained unsolved. After land reform, however, the "model" approach was given full scope; today it plays a major role in agricultural modernization. The "net and points" (wang-tien) method is to build up relatively successful agricultural units; then those units are joined with less advanced ones in a communications network, encouraging and demonstrating success for others to follow. Prominent components of this "net and points" effort are the "basic points" (chi-tien) of the provincial agricultural research institutes (nung-yeh k'o-hsueh-yuan). These "points" are actually experiment stations attached to selected brigades throughout each province, selected to exemplify particular soil or other environmental conditions. Attached to these stations are scientists from the institutes, sent down in rotation; there are also "peasant technicians" trained by these scientists to carry out experimental work. At any time, one-third to one-half of the scientists at any institute may be serving at the basic points. The scope of this extension work is impressive: the Shensi Provincial Agricultural

- 163 -

Institute, for instance, maintains contact with 100 basic points in 75 counties throughout the province and staffs each one with scientific personnel. Much of the work of these basic point stations is designed to discover which varieties will do best in a particular area. Other work, though termed experimental, is actually demonstrational; for instance, plantings of improved seeds next to other varieties in order to show peasants the advantages of the new over the old. Much of the experimental work said to be going on at the brigade level must be understood in this sense. The permanent linkage of research institutes to the model and net and points systems, however, is an extremely significant development for China's agriculture. Emulation remains an important trait in the culture, and China's administrative revolution now makes it possible to employ this trait in a systematic manner.

Revolution and Development

The key duality built into China's current political programs is exemplified by the slogan (seen everywhere in China) "grasp revolution, boost production." Many programs with a highly political coloration, presented in the revolutionary style, are best understood as combining two aspects: the effort to strengthen the Party's class-based political control, and the effort to promote those goals usually associated with modernization and economic development. For example, educational reform: the educational style that emerged from the Cultural Revolution emphasized a noncompetitive and nonelitist atmosphere in the classroom, with mutual aid among students, and a less authoritarian, elitist style for faculty. Among the innovations were open-book exams in which students were encouraged to remedy deficiencies rather than subjected to arbitrary standards and sanctions. In one sense, this aims at avoiding the development of an educational elite and promoting the chances of worker and peasant students; in another, however, it can be seen as a battle in the old campaign against stereotyped, rote learning, traditionally a bane of Chinese education. Thus besides being a politically important reorientation of classroom procedure, it also represents a modernization of education in line with trends elsewhere in the world.

Probably the clearest example of this link between revolution and development, and the one that most directly affects agriculture, is the "sending down" (hsia-fang) process. Basically there are three types of sending down: the movement of large numbers of high school or junior high school graduates to the villages, most of them permanently; the 30 days of manual labor expected of all cadres annually (in the sense of brain-workers); and longer term stints on farms or in factories for all brain-workers. This may involve as much as six months every two years, and in most cases probably requires actually living in the village or a "May Seventh" cadre school. This last is a major factor in every research and educational unit, as is described elsewhere in this report.

The revolutionary aspect of these types of send-down, particularly the first and the third, is the reeducation of the intellectuals by the workers and peasants and the consequent reorientation of their political attitudes. The development aspect is the injection of modern educated city talent into the under-developed world of the villages. If one asks which of these aspects is the more important, one is told that while both are important, the second clearly takes primacy. Marxism, one cadre explained, aims at the transformation of the objective world. This was a goal, whereas the transformation of thought is merely a method for attaining that goal (the means and ends relationship embedded in his statement is characteristically Maoist).

Historically, the social and cultural division between city and country-side has been a serious problem in China. Under the empire, the concentration of administrative functions and political influence in the walled cities proved a magnet for the literate elite. This gravitation of literate talent toward cities was facilitated since the 16th century by the increasing monetization of China's rural economy, which fostered absentee landlordism and tended to separate the elite from any direct concern for the management of agriculture or the fate of village society. This process intensified as China's coastal cities were exposed to Western commercial enterprise and Western values. Educated youth brought up in the modernized schools of these cities had little understanding of China's rural problems and (with a few notable exceptions, including some of China's present leaders) little inclination to learn about them. The result was the progressive loss of educated people in the villages who could respond effectively to the leadership needs of rural society. Though there were many other causes, we can probably attribute a good share of China's rural poverty to this dearth of educated and socially conscious leadership.

Today the Chinese express this historical problem theoretically in terms of "three great contradictions": between city and country, mental and manual labor, and worker and peasant. It is said that sending-down will ultimately help to resolve all three. In the short term, however, it is the first that is most directly affected, and the goals of rural economic development most directly served.

In fact, it appears that the educated middle-school youth are already making a significant impact upon China's rural leadership problem. Peasant technicians that we met in communes were sometimes middle-school graduates sent down from cities (and sometimes of local peasant origin). They were playing a prominent role in the experimental stations at brigade basic points. Some sent-down middle-school graduates had already assumed positions of local leadership as members of revolutionary committees of brigades. We were told that this phenomenon is quite common, and that their abilities are taken advantage of by local communities. Intellectuals who are sent down for six-month or year-long stays (we were told by one cadre who had experienced it) contribute toward educational and political work, promoting basic literacy, and helping with political education and the preparation of tatzupao (big-character posters). Though they adopt a low posture in public, scientists from the agricultural institutes who are sent down to basic point brigades in rotation are obviously playing a major role in the breeding and cultivation work of the experiment stations. All this suggests that the significance of the send-down program for rural development is very great, and that China's revolutionary political momentum is feeding directly into the drive to modernize village life and agriculture.

Nationalism and Internationalism

In assessing the future of China's relationship with the international scientific community, we are immediately confronted with the PRC's much advertised drive to achieve self-sufficiency in all major sectors of economic and intellectual life. China's modern historical experience has reinforced the determination not to become dependent upon the outside world for the fundamental requisites of national strength. The Western powers during the 19th and early 20th century used consortium loans to fasten a stranglehold upon key sectors of China's economy. More recently, the jolting experience with the Soviet Union has strengthened China's resolve for self-sufficiency.

History's practical lessons are naturally reinforced by feelings of national pride: justified pride in the talents of China's own scientists, and in the scientific contributions of China's own cultural heritage (exemplified particularly in the vigorous support of Chinese medicine). These feelings are given additional power by China's long history of great cultural achievements. The Cultural Revolution seems to have strengthened the nativist strain in contemporary Chinese thought in a number of areas, and it is at least possible that it has had some effect on scientific research in agricultural fields. Self-sufficiency slogans are prominently displayed on the walls of scientific research

- 166 -

institutes. This could have an entirely healthy effect upon Chinese science, emphasizing the need to build a strong, independent base in fundamental research. Though it is not at all clear how the self-sufficiency drive will affect China's willingness to participate in international scientific cooperation in agriculture, we were encouraged by scientists' evident interest in work being done at the various international centers. This suggests that China's scientific community realizes the interdependency of a strong domestic scientific base and regular participation in the international scientific community.

Yet we sensed a certain contradiction in China's current management of science. There is the entirely understandable drive for scientific self-sufficiency. There is also, however, the program of decentralization of research centers, an emphasis upon applied rather than basic work, and the interruption of basic laboratory study by the transfer of highly trained personnel to the country-side. Whether these two processes are really in basic contradiction is very difficult for the outsider to discern. We were aware, though, of the importance of the balance which China eventually strikes between these two components of agricultural policy.

Appendix 1

COMMITTEE ON SCHOLARLY COMMUNICATION
WITH THE PEOPLE'S REPUBLIC OF CHINA

PLANT STUDIES GROUP

RICHARD L. BERNARD
Research Geneticist
Agricultural Research Service
U.S. Department of Agriculture
U.S. Regional Soybean Laboratory
Urbana, Illinois 61801

Dr. Bernard's main interests are in the development of soybean varieties for commercial production with improved yield and pest resistance, in soybean germ plasm collection, and in the genetics of natural-occurring variations.

NORMAN E. BORLAUG
Director, International Wheat
  Improvement Program
International Maize and Wheat
  Improvement Center
Londres 40, Mexico 6, D.F.
Mexico

Dr. Borlaug, who won the Nobel Peace Prize in 1970, has strong interests in plant breeding methodology, improvement of wheat, barley, triticales, plant pathology and forest pathology, inter-generic crosses, and methods for rapidly increasing yields of food crops.

NYLE C. BRADY
Director, International Rice
  Research Institute
Los Baños, Laguna
Manila, Philippines

Dr. Brady is interested in rice breeding and culture, plant nutrition, research on soils and soil fertility, water management, multiple-cropping, and international cooperation in rice improvement.

GLENN W. BURTON
Research Geneticist and Leader
ARS, USDA, and
University of Georgia
Tifton, Georgia 31794

Dr. Burton's research interests center on the breeding, genetics, and management of forage and turf grasses, on animal production (especially pastures and forages), and millet breeding.

JOHN L. CREECH, Vice Chairman
Director
U.S. National Arboretum
Washington, D.C. 20002

Dr. Creech has conducted eight major plant explorations, mainly in Asia; his research interests lie in plant introduction and the exchange of genetic resources, and in the distribution of woody plant species of China and Japan.

JACK R. HARLAN
Crop Evolution Laboratory
Department of Agronomy
University of Illinois
Urbana, Illinois 61801

Dr. Harlan is interested in the origins of agriculture and cultivated plants, and the genetic relationships of cereals and their wild relatives. He is also concerned with taxonomic classification of sorghum and millets.

ARTHUR KELMAN
Professor and Chairman
Department of Plant Pathology
1630 Linden Drive
University of Wisconsin
Madison, Wisconsin 53706

Dr. Kelman is particularly concerned with bacterial diseases, the nature of resistance to bacterial pathogens, and possible relationships between Chinese scientists and the International Society of Plant Pathology.

HENRY M. MUNGER
Departments of Plant Breeding
  and Vegetable Crops
New York State College of
  Agriculture and Life Sciences
Cornell University
Ithaca, New York 14850

Dr. Munger is a horticultural scientist. Disease resistance in vegetables, hybrid vegetables, and nutrient production by vegetables are his special interests.

GEORGE F. SPRAGUE
Distinguished Professor
Department of Agronomy
University of Illinois
Urbana, Illinois 61801

Dr. Sprague's special interests include basic genetics and plant breeding methodology, with particular emphasis on maize and sorghum breeding and production.

STERLING WORTMAN, Chairman
Vice President
The Rockefeller Foundation
111 West 50th Street
New York, New York 10020

Dr. Wortman is interested in genetic improvement of rice, maize, and other basic food plants, in exchange of germ plasm, and in arrangements for international cooperation in agricultural research.

ALEXANDER P. DEANGELIS
Professional Associate
Committee on Scholarly Communication with the People's Republic of China
National Academy of Sciences
2101 Constitution Avenue, N.W.
Washington, D.C. 20418

Mr. DeAngelis, Secretary and translator for the Committee, specializes in the study of Chinese literature, especially folk literature. He is arranging visits of some delegations from the People's Republic of China.

PHILIP A. KUHN
Professor of Chinese History
The University of Chicago
Chicago, Illinois 60637

Dr. Kuhn, Asian Specialist and translator for the Committee, is a historian specializing in the history of China from 1644 to the present, including the history of Chinese political thought.

PROFESSIONAL INTERESTS OF PLANT STUDIES DELEGATION
TO THE PEOPLE'S REPUBLIC OF CHINA - MAY 7, 1974

The Plant Studies Delegation of the Committee on Scholarly Communication
with the People's Republic of China greatly appreciates the opportunity to visit
the People's Republic of China during the period August 27 to September 23, 1974.
We look forward to mutually useful, interesting, and friendly exchanges of views,
information, and materials.

The U.S. delegation comprises individuals representative of major areas
of plant studies. Each has many years of experience both in the development of
scientific theory and in its practical application to improvement of crop productivity on farms. The members are those who are accustomed to working long
hours in the field, and who prefer - as we suspect our Chinese counterparts do -
to see and discuss experiments in the field or laboratory to the maximum extent
possible. Consequently, the delegation is hoping to be privileged to visit
centers of field experimentation in several of the major agricultural regions
of China, and that its hosts will emphasize discussions in the field or laboratory rather than around the conference table.

Most members of the delegation met in Washington, D. C. on May 6-7 to
consider their mission and to prepare for their hosts in the People's Republic of
China a list of suggestions of institutions which might be visited, issues which
could be discussed, and materials which the delegation hopes may be seen. The
delegation trusts that its host, the Chinese Society of Agronomy, will find
these suggestions helpful. The delegation realizes that its suggestions are
based on inadequate knowledge of scientific institutions and activities of the
People's Republic of China, and that it must leave to its hosts the planning of
activities and visits which will make the visit useful to the scientists and
people of both nations.

The delegation hopes to accomplish four major objectives during its
four-week study of Chinese agriculture.

1. To learn about the nature and organization of Chinese agricultural sciences, including more basic as well as applied aspects of crop improvement, and to share its experiences with Chinese authorities.

2. To observe in the field the rich array of germ plasm of China's important crops, and to discuss prospects for mutually advantageous exchange of germ plasm. Also, to identify ways in which scientists of the United States and the People's Republic of China might work together - and with scientists elsewhere in the world - in the collection, evaluation, preservation, and exchange of germ plasm which is so important to all countries.

3. To develop an understanding of the agriculture, particularly crop production, in several important agricultural regions. This would require direct observation of farming practices and discussions with Chinese authorities of ways in which crop yields can be increased - how new, higher levels of productivity can be achieved.

4. To identify, if possible, opportunities for further cooperation in agricultural and related sciences, including more fundamental work, going beyond the exchanges of 1974. Such opportunities might include: more exchanges of scientific group visits; arrangements for joint conferences on scientific topics of mutual interest; research by individual scientists of one country at institutions of the other; collaboration on problems of an international scale; and improved exchange of technical information. Discussions of such possibilities - even if informal, as talks must be at this time - would be interesting and could be helpful to both nations.

The delegation, because of its large size and diverse interests, hopes that arrangements can be made for division of the delegation into two or more subgroups for visits to scientific and technical institutions or to farms in each region. This would allow the delegation to learn much more about China's efforts in the plant sciences, and would give smaller groups of Chinese and American specialists opportunities to discuss scientific topics of particular interest to them. The delegation hopes that it can divide into subgroups according to their particular interests to engage in discussions or field trips, whenever possible, and with similar small groups would visit different departments of an institution or different institutions in a city or region. The delegation hopes that its hosts will consider such arrangements wherever desirable and feasible from the hosts' viewpoint.

Scientific Interests

The delegation is especially interested in visiting major centers of field and laboratory experimentation with major crops from the northeast to the southern regions. Members of the delegation have indicated in the following paragraphs some of their special interests.

Germ Plasm

The delegation realizes that a vast reservoir of plant genetic resources exists in the People's Republic of China. Many of the world's most important food, feed, and fiber crops originated in mainland Asia and are cultivated by the Chinese people as well as elsewhere in the world.

It is the desire of the delegation to view current research on a broad array of crops, particularly those which are the primary food species in world agriculture. In addition, the delegation would like to observe production practices which have been developed successfully in China, as well as varieties selected for particular purposes.

We would like to observe the methods used in the People's Republic of China for the exploration, evaluation, and multiplication of native genetic resources. In addition, there would be great interest in the manner by which such germ plasm is maintained for future use.

The delegation is prepared to discuss informally the benefits of exchange of genetic resources and the mechanism whereby the genetic resources maintained by American scientists can be exchanged with their Chinese counterparts. By providing for the free exchange of wild, primitive, and advanced crop lines, the delegation believes that great strides can be made to expand crop yields, improve quality and adaptation, and improve resistance to pests. Equally important will be the provision for documentation of the materials being exchanged.

It is understood that in the People's Republic of China plant explorations are conducted yearly to collect native plant materials. We would like to know the organization of such activities, responsible groups, and methods of increase and dissemination of the plant materials procured.

There is great interest among American scientists in crops with new and useful constituents, alternate sources of fibers, and plants useful for essential oils, medicines, and industrial compounds. The American botanical community is most anxious to collaborate in field explorations directed toward a better understanding of the affinities of U.S. and Chinese flora. Indeed, joint efforts in this field will help to clarify the taxonomic and evolutionary relationships of genera having species native to both countries. Exchanges of herbarium materials are of scientific importance. The exchange of woody plant species useful for forestry, conservation of the land, and environmental horticulture to improve the welfare of man in urban areas would be beneficial.

The delegation plans to request permission to obtain small seed samples from the various localities visited. These will most likely be either wild species or old landraces that would be of scientific interest to American plant breeders. The delegation will want to discuss this issue at the first technical meeting with our Chinese hosts in Peking. For example, Dr. Bernard is most interested in the wild soybean and its relatives that might be encountered on the trip.

At the same time, the delegation plans to bring with it scientific samples of crop genetic materials that we believe would be of interest to Chinese plant scientists. If the delegation could learn in advance some of the materials that would interest Chinese scientists, we would attempt to bring them with us or make arrangements for sending them later.

Maize, Sorghum, Wheat, Rice and Millets

These are important crops in many areas of the People's Republic of China. Topics of general interest would include farming systems and recommended production practices; breeding methods employed; the extent of the use of varieties, synthetics or hybrids (and types of hybrids) used; methods of seed production and distribution; and methods of disease and insect control and breeding for resistance. Delegation members with such interest are Drs. Sprague, Wortman, Harlan, Borlaug, Brady, and Burton.

Forage, Pasture Grasses, and Legumes

Chinese improvement and management of grasses and legumes for forage, range and turf will be of interest to several members of the delegation. Exchange of experience with breeding, screening, and evaluation methods, including assessment with animals, will be desirable. Several delegation members would like to see some of the livestock (cattle, pigs, goats, etc.) that consume forage and pasture in China. Species of special interest will include the millets (Pennisetum, Panicum, and Setaria spp.), Zoysia spp., Cynodon spp., Eremochloa ophuroides, and Lespedeza spp. Dr. Burton would like to collect a few seeds or rhizomes of these species (particularly as they occur in the wild) if this can be arranged.

Horticultural Crops

Members of the delegation, particularly Drs. Munger and Creech, would be interested in visiting leading centers where research on vegetables is conducted. Vegetables in our terminology include potatoes, sweet potatoes, beans and peas, and melons as well as the usual leafy vegetables, root and bulb vegetables, the solanaceous vegetables, and the cucurbits.

They would wish to visit vegetable production areas to observe current production practices, varieties, pest control, and especially multiple cropping and intercropping of vegetables with each other with field crops.

They would also wish to observe the role of various agencies in the breeding and seed production of vegetables. Of special interest would be the breeding and seed production of F_1 hybrids, the extent to which seed producing agencies also do variety development, the system of distributing vegetable seeds, and the extent of seed selection and seed production by vegetable growers themselves.

We would wish to visit one or more centers of research on nutrition to obtain information about the nutrients contributed by vegetables to the Chinese diet, the planning for adequate production of specific nutrients, and educational programs on the use of various foods to produce balanced diets. Delegation members would also wish to observe fruit breeding whenever and wherever convenient.

Plant Pathology

The members of the delegation interested in plant pathology, particularly Drs. Kelman and Borlaug, would like to evaluate the methodology that has been developed for rapid and intensive screening of plant material for resistance to disease; and to determine the status of research in the areas of biological control of plant pathogens (soil-borne) by cultural practices, bacterial diseases of plants in general and in particular the pathogens affecting cereal grains, soybeans and vegetable crops (potatoes), the ecology of soft rot bacteria-survival and transmission by insects (Erwinia and Pseudomonas), the bacterial wilt caused by Pseudomonas solanacearum, the biochemical basis for disease resistance in plants, and the use of cell and tissue culture systems in studies on disease resistance. They also wish to become acquainted with the organization, research, extension, and instruction in plant pathology in relation to other agricultural sciences and the scientific societies that represent microbiologists and plant pathologists in the People's Republic of China.

Soybeans

Since China is the original homeland of the soybean, the observation of soybean research, production practices and problems, and especially of varieties and breeding methods will be of great interest and value. Observation in the field and description of the following are of especial interest to Drs. Bernard, Harlan, and Kelman

1. Production practices, including planting methods, density of population, tillage, cultivation, and harvest methods.

2. Soybean varieties grown commercially in each area, breeding methods, and objectives in variety improvement.

3. Pest problems and methods used in coping with them, including identification or description of major disease, insect, nematode, and weed pests; methods of control; and breeding for resistance.

4. Utilization of the soybean and breeding for improved utility of the seeds with respect to protein, oil, amino acid, and fatty acid composition.

5. Soybean germ plasm collections with main interest in sources and availability of the ancient native Chinese varieties as well as currently used commercial varieties; in sources of resistance to specific disease, insect, and nematode pests; and in any useful or unusual variation in growth type, drought resistance, temperature tolerance, etc. We maintain a germ plasm collection of about 3,000 strains in Illinois, many of them obtained in eastern Asia during the past 60 years. We are interested in enlarging this collection as a basis for future breeding efforts in the United States. We would be glad to arrange a reciprocal exchange of this material for Chinese germ plasm.

6. The wild soybean in its natural habitat and any research being done on its occurrence, diversity, use in breeding, and its ancestry to the cultivated soybean. We have a similar interest in the semi-wild or weedy soybeans of China. Because it is a great potential gene pool for pest resistance and perhaps other traits of breeding value, we would like to take a few seeds form wild plants that may be encountered in our trip, although we may be too early in the season for ripe seeds.

7. We would like to collect a few seeds from individual soybean fields and obtain seeds from breeders' collections whenever possible. A collection of bacterial modules from various fields would be of interest and value to our microbiologists studying N-fixation in soybeans.

8. Since soybeans are grown widely throughout eastern China and as far west as Inner Mongolia, Ch'ing-hai, and Yunnan, almost anywhere in China will be of interest. We would like to see soybean breeding centers in each area visited. Soybean production is most intense in the north of China and our major interest is there. Yet our germ plasm and our knowledge is most deficient in the south and west, and we are most anxious to learn about and obtain the diverse soybean varieties from these areas.

Origin and Evolution of Cereal Crops

The origin and evolution of cereal crops with special reference to their wild relatives is of particular interest to Dr. Harlan. His specialty is with sorghum and millets, but he is interested in rice and other cereals and the broad field of genetic resource management as well. He also has an interest in archeobotany of the neolithic period with respect to the origins of agriculture, and he would be interested in speaking with Chinese archeologists on this subject.

Forestry

The delegation would be interested, as the opportunity permits, to observe forestry activities as they pertain to land reclamation, trials with species for conservation of soil and water, and for protection of croplands against environmental stress. Forestry nursery practices would be interesting to observe when it would coincide with other delegation objectives.

Wide Genetic Crosses

Several members of the delegation are interested in intergeneric crosses, cell and tissue culture, somatic hybridization, and related genetic advances useful in plant breeding. Drs. Borlaug, Burton, and Wortman are especially interested in these topics.

Suggested Organization of the Visit

The delegation believes, on the basis of available information, that a series of visits to major regions of the People's Republic of China would be highly desirable. The delegation suggests a schedule approximately as follows:

<u>August 27</u> - Arrive from Hong Kong; overnight at Kwangchou

<u>August 28</u> - Travel to Peking

<u>August 29 - September 2</u> - In the Peking area (5 days)

The delegation would welcome an opportunity to become acquainted with leaders and learn about the activities in plant sciences of the Chinese Academy of Sciences, the Chinese Academy of Agricultural and Forestry Sciences, and the Ministry of Food and Agriculture.

Members of the delegation have special interests in visits to the Institute of Botany (full day), the Institute of Applied Mycology, the Institute of Plant Taxonomy, the Botanical Garden of Peking, the North China Institute of Agricultural Research, and a vegetable production commune at which agricultural experimentation is an important activity. We have heard that there also exists in Peking a Research Laboratory of Agricultural Heredity which could also be of interest.

Arrangements might be made in Peking for discussions of China's activities in plant germ plasm collection, evaluation, and preservation.

Dr. Kelman wishes especially to meet the leaders of the Chinese Society of Plant Pathology and the Chinese Society of Microbiology. Dr. Munger hopes to meet leaders of the Chinese Society of Horticulture.

The delegation would like to meet with its host organization, the Chinese Society of Agronomy.

<u>September 3</u> - Travel to Kirin (Jilin) Province

- 177 -

September 4-6 - In Kirin Province

Visits would be appreciated to centers of experimentation with and production of soybeans, maize, sorghum, millets, and other major crops. Among institutions that might be visited are the Northeast China Institute of Agricultural Science at Kung-chu-ling and a commune where experimentation in crop production is important.

September 7 - Travel to Sian and Ch'eng-tu, or split the team

September 8-11 - In Shensi Province

The delegation would like to spend time both in the Sian and Ch'eng-tu areas, if possible. Or, half of the delegation could go to Sian, and half to Ch'eng-tu.

In Sian, the delegation is interested in activities of the Northwest Institute of Agricultural Sciences, the Institute of Pomology (some members), the Botanical Garden of Sian, and other institutions doing important work in the plant sciences. Observation on farms of crops and cropping systems would be desirable.

The delegation has heard of agricultural activities at Tachai Commune, but does not know if time and circumstances would permit a stop there: this matter is left to the discretion of the hosts.

The Szechwan Basin is reportedly one of the most productive agricultural regions of China. It is a region of long cropping seasons capable, perhaps, of producing most of the crops grown elsewhere in China. We believe that it would be an area in which to observe the highest production efficiency in crops such as wheat, corn, rape, sweet and white potatoes, fruits, and vegetables.

Therefore we propose that during the period of September 7-11, the delegation visit Ch'eng-tu as an optional alternative to the city of Sian, or that the delegation split into two groups at this point. One team would go to Sian, and the other to Ch'eng-tu. The two teams would rejoin in Nanking on September 12. The most desirable arrangement, of course, would be for the full delegation to visit both Sian and Ch'eng-tu for shorter periods.

September 12 - Travel to Nanking or Shanghai

September 13-17 - Nanking, Shanghai, Suchou area

It is understood that there are many important agricultural institutions and activities in this region. Among those that seem to be of major interest are the Shanghai Academy of Agricultural Science and the Institute of Biochemistry, the Institute of Plant Physiology, the Institute of Horticulture, the Soil Fertilizer and Plant Protection Institute, and the Sun Yat-sen (Sun Zhong-shan) Botanical

Garden of Nanking. The delegation probably should be split into at least two
teams for visits in this region, as there appears to be a number of interesting
and important places. The feasibility of a visit of some members to the Botanical
Garden of South China, Hangchow, Chekiang, might be considered by our hosts. Also,
Dr. Kelman would like to meet leaders of the Phytopathological Society of China.
September 18 - Travel to Kwangtung Province (Probably Kwangchou)
September 23 - Depart from Hong Kong

        In presenting to the Chinese Society of Agronomy the above suggestions
of scientific interests, institutions to be visited, issues to be discussed, and
dates for the visits to various organizations, the delegation wishes to express
its hope that the Society will make such additions, deletions or changes as it
feels would improve the usefulness of the trip.

ACTUAL ITINERARY

August 25
Sunday

Members of the delegation met for dinner in Kowloon, Hong Kong. Final details of the trip were discussed, as were writing assignments of each member for the team's report.

August 26
Monday

Briefing at the U.S. Consulate by Consul-General Charles Cross, and by Messrs. Sherrod McCall, Wever Gim, Linwood Starbird, and Harold Champeau (Agricultural Attache). Train tickets were purchased for the trip to the border at Lo Wu, team members picked up publications and seed samples which had been forwarded to the Agricultural Attache, and last minute purchases were made.

August 27
Tuesday

Left Kowloon by train.
Arrived Lo Wu, Hong Kong. Crossed border on foot to Shumchun, People's Republic of China. Received by Mr. Li Chin-ch'ang of the Association of Agriculture, Kwangtung Province.
Lunch
Boarded train for Canton.
Arrived Canton. Received by Mr. Liu Liang-jung of the Kwangtung Association of Agriculture. Tour of the Kwangchou (Canton) City Museum. Overnight at the Tungfang Hotel, Canton.

August 28
Wednesday

Visit to White Cloud Mountain Park followed by a roadside stop at an unidentified commune where vegetable plots and paddy rice experiments were seen.
Left for airport for an 1800 flight to Peking.
Arrived Peking. Met by Mr. Ma Ling, Deputy Secretary-General of the Chinese Association of Agriculture, Professor Lou Ch'eng-hou, and others.
Checked in at the Peking Hotel.

August 29
Thursday

Meeting with Professor Lou and others to discuss and agree on itinerary.
Imperial Palaces.
Nan-yuan Peoples Commune.
Banquet given by Mr. Ma on behalf of the Association of Agriculture.

August 30
Friday

Academy of Agricultural and Forestry Sciences.
Summer Palace.
Atomic Energy Utilization Research Institute.
Performance of Marshal Arts.

August 31
Saturday

Academia Sinica. Received by Professor Wu You-hsun.
Institute of Botany, Academia Sinica.
Institute of Genetics, Academia Sinica.
Presentation of Chinese films.

September 1
Sunday

Visits to Ming Tombs and Great Wall, with three roadside stops
for examination of fields of corn and sorghum; millets and corn;
and cotton.
Banquet hosted by U.S. Plant Studies Delegation.

September 2
Monday

Visit to produce market, Peking.
Presentation by the U.S. Delegation to the Chinese Association
of Agriculture of publications and germ plasm from U.S. scientists
and institutions.
Discussion groups centered on wheat and rice; corn and sorghum;
vegetables and soybeans; germ plasm exchange; and plant protection.
Boarded train for Kung-chu-ling, Kirin Province.

September 3
Tuesday

Arrived Kung-chu-ling for visit to Kirin Academy of Agricultural Sciences (KAAS).
Presentation on organization, activities and history of KAAS,
by Vice President Li Yi.
Examination of field experiments with Setaria millets, corn,
soybeans, sorghum, cropping systems.
Banquet hosted by Vice President Li Yi.

September 4
Wednesday

Big Elm Tree Production Brigade. Visit to field plots involving
inbreds and double crosses of corn, sorghum, in cooperation with
KAAS.
Paddy rice experimental plots, KAAS.
Discussion groups, centering on wheat and rice (Borlaug, Brady);
corn, sorghum, and millets (Sprague, Harlan, Burton); plant pro-
tection (Kelman); soybeans (Bernard); and visit to Institute of
Fruit Breeding (Creech, Munger, DeAngelis).
Film on Chinese acrobatics at Academy theatre.

September 5
Thursday

Primary school, KAAS.
KAAS presentation of 40 seed samples to U.S. Delegation.
Depart for Ch'ang-ch'un, capital of Kirin Province.
Association of Agriculture, Kirin Province. Received by Mr. T'eng
Wen, Vice President.
Luncheon hosted by Mr. T'eng.
Departure for Peking on aircraft of Chinese Civil Aviation Admin-
istration (CAA).
Arrived Mukden (Liaonang). Unable to proceed because of poor
visibility in Peking. Overnight at Friendship Hotel, as guests
of the Chinese Association for Friendship with Foreign Countries
(CAFFC) and the Association of Agriculture.
Dinner hosted by Mr. Wang Wen-kuei, Shenyang Branch, CAFFC and
Mr. Wang Shou-ye, Liaoning Branch, Association of Agriculture.

September 6
Friday

Departed for Peking on CAA plane.
Arrived Peking. To Peking Hotel.
Meeting with the Acting President, Chinese Association of Agri-
culture, Mr. Hao Chung-shih.
Reception, U.S. Liaison Office. Hosts: Mr and Mrs. Bruce.
Departed by train for Sian, Shensi Province.

September 7
Saturday

Arrived Sian. Met by Mr. Tu Lu-kung, Deputy Director, Shensi
Council of the Association of Agriculture and Mr. Liu Hsien-chen,
Chairman of the Revolutionary Committee of the Academy of Agricultural and Forestry Sciences of Shensi Province. Arranged
itinerary for Sian.

September 8
Sunday

Two groups departed by car for SAAFS and Fruit Research Insti-
tute.
Visited Academy of Agricultural and Forestry Sciences, located
in Wu-kung County and the Fruit Research Institute in Mei
County (Creech, Munger, DeAngelis).
Banquet hosted by Mr. Tu Lu-kung.

September 9
Monday

Visits to Hua-ch'ing Hot Spring and Pan-p'o Museum.
Red Star Production Brigade (vegetables). Bell Tower.

September 10
Tuesday

Northwest College of Agriculture.

September 11
Wednesday

Presentation of seed samples by the Association of Agriculture
of Shensi Province.
Departed by train for Nanking.

September 12
Thursday

Arrived Nanking.
Visit to zoo and parks, and to Yangtze River Bridge.
Banquet given by the Kiangsu Province Association of Agriculture,
Mr. Ku T'ing, Chairman.

September 13
Friday

Kiangsu Academy of Agricultural and Forestry Sciences, Nanking.
Visit to Tomb of Dr. Sun Yat-sen, followed by discussions in
groups with Chinese scientists from the Academy.
Soils Research Institute of Academia Sinica.

September 14
Saturday

Presentation of seed by Mr. Ku T'ing, Chairman, Kiangsu Asso-
ciation of Agriculture.
Departed by train for Shanghai.
Arrived Shanghai. Met by Mr. Hsiao You-shen, Deputy Director
of the Council, Shanghai Association of Agriculture, and Mr.
Chen Yu-hua, Deputy Secretary-General.
Sightseeing.
Opera, "Azalea Mountain."

September 15
Sunday

Children's Palace.
Hung-Ch'iao (Rainbow Bridge) Peoples Commune.

September 16
Monday

Lung-hua Tree Nursery, and exhibit of P'en-ching (dwarfed)
plants.
Shanghai Academy of Agricultural Sciences.
Banquet given by Mr. Hsiao You-chen, Shanghai Association of
Agriculture.

September 17
Tuesday

Visit to Shanghai market (Munger, Creech, Kelman, Burton, Wortman).
Shanghai Electric Machine Company.
Institute of Plant Physiology, Shanghai Municipality.
Institute of Biochemistry, Academia Sinica.

September 18
Wednesday

Industrial Exhibition, Shanghai.
Presentation of seed samples by Association of Agriculture of Shanghai.
Departed by plane for Canton.
Arrived Canton. Met by Mr. Liu Liang-jung, Vice Chairman, Kwangtung Association of Agriculture, and by Mr. Li Chin-chang.
Conference on itinerary.

September 19
Thursday

Kwangtung Botanical Garden.
(a) Chungshan (Sun Yat-sen) University, and its herbarium (entire group except Brady, Kelman, Wortman).
(b) Grain Crops Research Institute (mostly rice, at Canton) of the Kwangtung Academy of Agricultural Sciences (Brady, Kelman, Wortman).

September 20
Friday

Hsin-ch'iao Peoples Commune (primarily rice).

September 21
Saturday

Lo-kang Peoples Commune (subtropical fruits).
Technical discussions with Chinese colleagues.

September 22
Sunday

Canton market.

September 23
Monday

Departed for Hong Kong.

September 24
Tuesday

Hong Kong. Drafting of report.

September 25
Wednesday

Hong Kong. Drafting of report.

Appendix 4

INDIVIDUALS MET IN CHINA

The following list of plant scientists and professionals in agriculture met during our visit is first arranged geographically according to the order of stops on our itinerary (omitting the first overnight stop in Canton on our way to Peking, and our overnight stop in Peking on our way to Sian). The order is as follows: Peking; Kirin Province, including the capital, Ch'ang-ch'un, and Kung-chu-ling in Huai-te County; an unscheduled overnight stop in Shenyang City, the capital of Lianoning Province; Shensi Province, including Sian City, Wu-kung County and Mei County; Nanking, the capital of Kiangsu Province; Shanghai Municipality; and Canton, the capital of Kwangtung Province.

Under each geographical heading we first list the Association of Agriculture of the People's Republic of China (Chinese Association of Agriculture), our host organization, then the other organizations alphabetically. Individual names are listed alphabetically. Asterisks are used to indicate the persons of responsibility and top scientists in each organization. As in the Chinese usage, surnames appear first. We met a great many people on a very busy tour schedule. Later, we were at times unable to match some of the names with people. In these instances no gender is indicated. Though we have attempted to record the names and titles carefully, there certainly are errors and omissions for which we apologize to our Chinese hosts.

The five people who traveled with us throughout China deserve special mention as it was largely due to them that our visit achieved its measure of success.

陈天林 Mr. Ch'en T'ien-lin: interpreter, Association of Agriculture and
                                                English teacher, Peking Foreign Language Institute

周振宁 Mr. Chou Chen-ning: Secretary of the Association of Agriculture

徐锦华 Ms. Hsu Chin-hua: interpreter, Association of Agriculture

黄永宁 Mr. Huang Yung-ning: Secretary of the Association of Agriculture

楼成后　　Professor Lou Ch'eng-hou:　Council Member of the Association of Agri-
　　　　　　　　　　　　　　　　　　　culture and Professor of Plant Physiology at
　　　　　　　　　　　　　　　　　　　North China (Hupei) College of Agriculture

PEKING

Association of Agriculture of the People's Republic of China (Chinese Association
　of Agriculture).

陈天林　　Mr. Ch'en T'ien-lin:　worker

周振宁　　Mr. Chou Chen-ning:　secretary

郝中士 *　Mr. Hao Chung-shih:　Acting Council Director, Association of Agriculture,
　　　　　　　　　　　　　　　　and Vice Minister of the Ministry of Agriculture and
　　　　　　　　　　　　　　　　Forestry

黄永宁　　Mr. Huang Yung-ning:　secretary

耿锡株 *　Mr. Keng Hsi-tung:　Council Member, also Responsible Person for the Office
　　　　　　　　　　　　　　　of the Chinese Academy of Agricultural and Forestry
　　　　　　　　　　　　　　　Sciences

梁勇 *　　Mr. Liang Yung:　Council Member

楼成后 *　Mr. Lou Ch'eng-hou:　Council Member, also Professor of Plant Physiology
　　　　　　　　　　　　　　　　at North China (Hupei) College of Agriculture

马凌 *　　Mr. Ma Ling:　Deputy Secretary General of the Association of Agriculture
　　　　　　　　　　　　and Vice Director of the Bureau of Foreign Affairs of the
　　　　　　　　　　　　Ministry of Agriculture and Forestry

孙基莲　　Ms. Sun Chi-lien:　worker

杨鸿　　　Ms. Yang Hung:　worker, interpreter

Association of Agriculture, Chinese Society of Forestry

吴中伦　　Mr. Wu Chung-lun:　Council Member: forestry

Chinese Academy of Agricultural and Forestry Sciences

章一华　　Mr. Chang Yi-hua:　plant pathologist

金善宝 *　Mr. Chin Shan-pao:　professor and one of the Responsible People of the
　　　　　　　　　　　　　　　Chinese Academy of Agricultural and Forestry Sciences,
　　　　　　　　　　　　　　　also Deputy Council Director of the Association of
　　　　　　　　　　　　　　　Agriculture; wheat specialist

邱式邦　　Mr. Ch'iu Shih-pang:　entomologist

李汉林 *　Mr. Li Hsi-lin:　Responsible Person for Scientific Research and Production,
　　　　　　　　　　　　　　also Council Member of the Association of Agriculture

- 186 -

林世成　　Mr. Li Shih-ch'eng:　professor and rice expert

苏格曼 *　Mr. Su Ko-man:　Responsible Person

王棠义　　Mr. Wang Ch'ung-yi:　Responsible Person for the Food Crops Research
　　　　　　　　　　　　　　　　　Laboratory working in octoploid triticale

王素　　　Ms. Wang Su:　vegetable expert

Chinese Academy of Agricultural and Forestry Sciences, Atomic Energy Utilization Research Institute

赵文楼 *　Mr. Chao Wen-p'u:　Responsible Person

Chinese Academy of Sciences (CAS)

陆思麟 *　Mr. Lu Ssu-lin:　Responsible Member

邓淑惠 *　Mr. Teng Shu-hui:　Responsible Member

吴有训 *　Mr. Wu You-hsun:　professor and Vice President; physicist

CAS - Institute of Botany

　　　　　Mr. Jen Hsu:　Director, Paleobotany Department

李森　　　Mr. Li Sen:　Responsible Person of the Revolutionary Committee

林榕 *　　Mr. Lin Ke:　Director:　plant taxonomist

崔澂　　　Mr. Ts'ui Ch'eng:　plant physiologist

CAS - Institute of Genetics

庄家骏　　Mr. Chuang Chia-chun:　wheat pollen culture, geneticist

庄巧生 *　Mr. Chuang Ch'iao-sheng:　senior wheat breeder

陈坚　　　Mr. Ch'en Chien:　wheat breeder

肖洪祐 *　Mr. Hsiao Hung-you:　Vice Chairman of the Revolutionary Committee

林建兴　　Mr. Lin Chien-hsing:　soybean breeder

马缘生　　Mr. Ma Yuan-sheng:　wheat breeder

鲍文奎 *　Mr. Pao Wen-k'uei:　genetic breeding, senior octoploid triticale breeder

曹道孝　　Mr. Tseng Tao-hsiao:　wheat breeder

董玉琛　　Ms. Tung Yu-chen:　wheat breeder

汪安琦　　Ms. Wang An-ch'i:　chromosome research

翁曼丽　　Ms. Weng Man-li:　microbiologist

Mr. Yeh Fan: potato breeder

尸福垚 Mr. Ying Fu-yu: wheat breeder

余产波 Mr. Yu Ch'an-p'o: heterosis utilization

Chinese Scientific and Technical Association (CSTA)

蒋琪 Ms. Chiang Ch'i: worker

朱永行 * Mr. Chu Yung-hsing: Deputy Director of the Foreign Affairs Office of the CSTA and CAS

李明德 Mr. Li Ming-te: worker

Ministry of Foreign Affairs - American and Oceanian Affairs Department

程香虹 Ms. Ch'eng Ch'i-hung: Deputy Director

余祖元 Ms. Yu Tzu-yuan: worker in the American and Western Hemisphere Section

Peking City Revolutionary Committee

史美煌 Mr. Shih Mei-huang: worker

KIRIN PROVINCE

Ch'ang-ch'un City

Association of Agriculture - Kirin Province Branch

潘顺法 Mr. Fan Sun-fa: worker

韩金 Mr. Han Chin: worker

谢炎 Mr. Hsieh Yen: wheat breeder

胡冬 * Mr. Hu Tung: Council Member

滕文 * Mr. T'eng Wen: Council Director

Kung-chu-ling City (Huai-te County)

Huai-te County Revolutionary Committee

石俊峰 Mr. Shih Chun-feng: Vice Chairman

Kirin Academy of Agricultural Sciences (KAAS)

隗莼 Mr. Huai Ch'un: Responsible Person for the Office (staff director)

李义 * Mr. Li Yi: Vice President

李义中 * Mr. Li Yi-chung: Vice President; insect pests

KAAS - Institute of Crop Breeding

張子金 Mr. Chang Tzu-chin: soybean specialist

金蓮花 Ms. Chin Lien-hua: millet specialist

谢道宏 Mr. Hsieh Tao-hung: corn specialist

Mr. Hsieh Yen: wheat breeder

顾模 Mr. Ku Mo: fruit tree specialist

侯连运 Mr. Hou Lien-yun: rice specialist

李仞 Mr. Li Ch'e: rice specialist

李公德* Mr. Li Kung-te: Deputy Director: sorghum breeder

王进先 Mr. Wang Chin-hsien: wheat breeder

吴洪文 Mr. Wu Hung-wen: rice specialist

KAAS - Institute of Plant Protection

胡吉成* Mr. Hu Chi-ch'eng: Director: plant pathologist

KAAS - Institute of Soil, Fertilizer and Cultivation

王产丰 Mr. Wang Ch'an-feng: soybean agronomist

武克忠 Mr. Wu K'e-chung: upland crops wheat agronomist

Mr. Yang Ching-chien: soil specialist

LIAONING PROVINCE

Shenyang City

Association of Agriculture - Liaoning Branch

王守业* Mr. Wang Shou-ye: Secretary General

Chinese Foreign Friendship Association - Shenyang Branch

王文贵 Mr. Wang Wen-kuei: Responsible Person

SHENSI PROVINCE

Sian

Association of Agriculture - Shensi Branch

安危 Mr. An Wei: worker

刘随仓 Mr. Liu Sui-ts'ang: worker

杜魯公 * Mr. Tu Lu-kung: Deputy Council Director

楊篤 * Mr. Yang Tu: Secretary General

Mei County

Shensi Province Academy of Agricultural and Forestry Sciences - Institute of
Fruit Research (2 1/2 hours by car from Sian)

陳策 Mr. Ch'en Ts'e: Director of the Plant Protection Laboratory

李世奎 Mr. Li Shih-k'uei: Director of the Cultivation Research Laboratory

林衍 Mr. Lin Yen: Responsible Person for Technology

崔紹良 Mr. Ts'ui Shao-liang: Deputy Director of the Breeding Research Laboratory

董甦 * Mr. Tung Su: Deputy Director of the Revolutionary Committee

王治謙 Mr. Wang Ju-ch'ien: technologist, cadre

Wu-kung County

Northwest College of Agriculture (2 1/2 hours by car from Sian)

常君常 Mr. Chang Chun-ch'ang: saline soils

張海峰 Mr. Chang Hai-feng: wheat breeder

趙洪璋 Mr. Chao Hung-chang: professor and chief wheat breeder

趙宜謙 Mr. Chao Yi-ch'ien: senior plant pathologist

程群力 Mr. Ch'eng Ch'un-li: worker-peasant-soldier student

荊家海 Mr. Ching Chia-hai: plant physiologist

朱德余 Chu Te-yu: Assistant Teacher, rice

馮立孝 * Mr. Feng Li-hsiao: President of College

何金江 Mr. Ho Chin-chiang: wheat breeder

耿志訓 Mr. Keng Chih-hsun: senior wheat breeder

李振岐 * Mr. Li Chen-ch'i: Associate Professor of Plant Pathology and Chairman
of the Department of Plant Protection

李振聲 Mr. Li Chen-sheng: Teacher, wheat breeder

李志才 * Mr. Li Chih-ts'ai: Deputy Director of the School Office

李洪元 Mr. Li Hung-yuan: Chairman of the Vegetable Teaching and Research Group

李生秀 Mr. Li Sheng-hsiu: geology

刘敬修 * Mr. Liu Ching-hsiu: Chairman of the Revolutionary Committee

刘孝谦 Mr. Liu Hsiao-ch'ien: Deputy Director of Teaching

刘太安 Liu T'ai-an: agronomist

路广明 * Mr. Lu Kuang-ming: Professor of Fruit Tree Cultivation, Chairman of the Department of Horticulture

毛绳绪 * Mr. Mao Sheng-hsu: Associate Professor of Forest Surveying, Chairman of the Department of Forestry

宋哲民 Mr. Sung Che-min: wheat breeder

宋玉迟 * Mr. Sung Yu ch'ih: Associate Professor of Corn Breeding, Vice Chairman of the Department of Agronomy

王建忠 Wang Chien-chung: soil fertility

王立祥 Wang Li-hsiang: rice

王 鸣 Wang Ming: Vegetable Teaching and Research Group, genetic breeding

吴守仁 Mr. Wu Shou-jen: professor

阳振民 Mr. Yang Chen-min: Vice Chairman of the Vegetable Teaching and Research Group

阎迺献 Mr. Yen Nai-you: Associate Professor of the Fruit Tree Teaching and Research Group

KIANGSU PROVINCE

Nanking

Association of Agriculture - Kiangsu Branch

    * Mr. Ku T'ing: Chairman of the Council

Mr. Ling Shih-hsin: worker

Mr. Mao Jen-liang: worker

Kiangsu Province Agricultural Sciences Research Institute

邹江有 Mr. Chou Chiang-you: Assistant Researcher, rice

周朝飞 Mr. Chou Ch'ao-fei: Assistant Researcher, wheat breeder

奚元龄 Mr. Hsi Yuan-ling: Researcher, physiologist, cotton breeder, genetic breeding

过从俭 Mr. Kuo Ts'ung-chien: Assistant Researcher, plant pathologist

沈梓培 Ms. Li Wei-fung: Researcher, Chinese cabbage

Mr. Shen Tzu-p'ei: Researcher, soils

Mr. Shu Hu-leng: Researcher, tomatoes

唐秀娟 Ms. T'ang Hsiu-chuan: Assistant Researcher, physiology, biochemistry

崔继林 Mr. Ts'ui Chi-lin: Associate Researcher, rice physiology

杜正文 Mr. Tu Cheng-wen: Responsible Person for the Plant Protection Research Group, Assistant Researcher, insect pests

吴光南 Mr. Wu Kuang-nan: Associate Researcher, rice physiologist

吴光远 Mr. Wu Kuang-yuan: Associate Researcher, vegetables, horticulturalist

杨致平 * Mr. Yang Chih-p'ing: Vice Chairman of the Revolutionary Committee

Ms. Yang Shu-ying: Researcher, vegetables in general

杨运生 * Mr. Yang Yun-sheng: Responsible Person for the Scientific Research Group, Associate Researcher, green manure and forage

阮德成 Mr. Yuan Te-ch'eng: Assistant Researcher, animal husbandry and feeds

Institute of Soil Research

陈万才 Mr. Ch'en Wan-ts'ai: Planning Worker

熊 毅 * Mr. Hsiung Yi: soil physics and chemistry, professor

共子同 Mr. Kung Tzu-t'ung: soil geography scientific worker

李庆逵 Mr. Li Ch'ing-k'uei: soil agronomy and chemistry, professor

姚贤良 Mr. Yao Hsien-liang: soil physics scientific worker

SHANGHAI

Association of Agriculture - Shanghai Branch

张光祖 Mr. Chang Kuang-tzu: worker

陈玉华 * Mr. Ch'en Yu-hua: Deputy Secretary General

周国祺 Mr. Chou Kuo-ch'i: technologist

周林挑 Mr. Chou Lin-t'iao: technologist

肖习深 * Mr. Hsiao You-shen: Vice Chairman of the Council

杨禹臣 Mr. Yang Yu-chien: worker

Lung-hua Nursery

        Mr. Shu Chung-kai: Director

Shanghai Academy of Agricultural Sciences (SAAS)

章振华   Ms. Chang Chen-hua: pollen culture

张国强   Mr. Chang Kuo-ch'iang: wheat breeder

卡女贞   Ms. Chia Nu-chen: corn technologist

储昕 *   Mr. Ch'u Hsin: Responsible Person for the Scientific Research Group

肖亮钧   Mr. Hsiao Liang-kou: Responsible Person for the Office

             Huang Pei-chung: spring barley breeder

刘日新   Liu Jih-hsin: rice cultivator

汤其戚 *   Mr. T'ang Ch'i-sheng: Vice Chairman of the Revolutionary Committee

蔡福根   Ts'ai Fu-ken: rice breeder

汪树俊   Wang Shu-chun: plant pathologist

SAAS - Institute of Horticultural Research

钟仲贤   Mr. Chung Chung-hsien: scientific and technical staff member, vegetables

许秀莲   Mr. Hsu Hsiu-lien: scientific and technical staff member, edible fungi

李培基   Mr. Li P'ei-chi: scientific and technical staff member, fruit trees

布炳根   Mr. Pu Ping-ken: scientific and technical staff member, vegetables

沈延松   Mr. Shen Yen-sung: scientific and technical staff member, vegetables

杨式琅 *   Mr. Yang Shih-lang: Vice Chairman of the Revolutionary Committee, horticulturalist

姚文莘   Mr. Yao Wen-ou: scientific and technical staff member, plant protection

Shanghai Institute of Biochemistry

冯宗铭   Mr. Feng Tzung-ming: Ribosome Laboratory specialist

黄爱珠 *   Mr. Huang Ai-chu: Responsible Person for Scientific Research

龚祖埙   Mr. Kung Tzu-hsun: specialist of the Virus Group of the Protein Laboratory

龚岳亭   Mr. Kung Yueh-t'ing: specialist in the Insulin Group of the Protein Laboratory

沈昭文   Mr. Shen Chao-wen: Professor of Biochemistry

王德宝　　Mr. Wang Te-pao: Professor of Biochemistry

武产荣 *　Mr. Wu Ch'an-jung: Chairman of the Revolutionary Committee

Shanghai Institute of Plant Physiology

张坤诚　　Mr. Chang K'un-ch'eng: Responsible Person for the Cell Physiology
　　　　　　　　　　　　　　　　　　Laboratory

陈敬祥　　Ch'en Ching-hsiang

陈永宾　　Mr. Chen Yung-ping: Responsible Person for the Nitrogen Fixation Group

郑幼霞　　Cheng You-hsia

沈镇德　　Mr. Shen Ch'i-te: Responsible Person for the Organ Shedding Group

沈允钢　　Mr. Shen Yun-kang: professor, photosynthesis

汤章城　　Mr. T'ang Chang-ch'eng: Responsible Person for the Phytotron

汤玉玮　　Mr. T'ang Yu-wei: Responsible Person for the Plant Hormone Research
　　　　　　　　　　　　　　　　Laboratory

王光远　　Wang Kuang-yuan: Cell Fusion Group

严国钩 *　Mr. Yen Kuo-kou: Vice Chairman of the Revolutionary Committee

殷宏章　　Mr. Yin Hung-chang: Responsible Person for the Photosynthesis Laboratory

KWANGTUNG

Association of Agriculture - Kwangtung Branch

姜志新　　Mr. Chiang Chih-hsin: worker

蒋景才　　Mr. Chiang Ching-ts'ai: worker

蒋才锦　　Mr. Chiang Ts'ai-chin: worker

韩宏光　　Mr. Han Hung-kuang: worker

李进昌 *　Mr. Li Chin-ch'ang: Council Member

刘良荣 *　Mr. Liu Liang-jung: Director of the Council

区灼辉　　Mr. Ou Ch'i-hui: worker

王慧玉　　Ms. Wang Hui-yu: interpreter

王淑贞　　Mr. Wang Shou-chen: worker

Canton Foreign Friendship Association

黄眈瀅 Ms. Huang Yang-teng: worker

Kwangtung Botanical Gardens

陈封怀 Mr. Ch'en Fang-huai: Professor, plant scientist

Mr. Kuan Hsueh-liang: Director of Forestry Laboratory

* Mr. Ts'ui Kuo-ch'eng: Director

Mr. Ye Yu-yen: technologist

Kwangtung College of Forestry

Ms. Chiang Ying: professor

Kwangtung Province College of Agriculture

范怀忠 Mr. Fan Huai-chung: professor, Vice Chairman of Department of Plant Protection

李沛文 Mr. Li P'ei-wen: fruit tree expert

Kwangtung Province Academy of Agricultural Sciences - Food Crops Research Institute

汪庆明 Mr. Chiang Ch'ing-ming: Vice Director, responsible for administrative affairs

谢豪 Mr. Hsieh Hao: Associate Researcher, responsible for rice cultivation

徐炎康 Mr. Hsu Yen-t'ang: Researcher, responsible for rice breeding

柯韋 Ms. K'e Wei: Researcher, responsible for rice breeding

Ms. Liu Shih-cheng: in charge of wheat breeding

伍尚忠 Mr. Wu Shang-chung: Associate Researcher, responsible for rice plant protection

Appendix 5

GERM PLASM PRESENTED AND RECEIVED

Germ Plasm Presented

The U.S. Plant Science Delegation brought a collection of seeds and plants to China for presentation to Chinese scientists. The main purpose was to demonstrate a willingness to begin an exchange of germ plasm. The materials provided a sampling of the broad array of germ plasm that is held in U.S. scientific institutions and the international institutes. It also was a means of showing the specificity of U.S. germ plasm in terms of sources of disease and insect resistance in which both wild species and advanced cultivars play important roles.

The materials presented to the Chinese Association of Agriculture and Forestry, Peking were as follows:

- 26 species of forest trees that totaled over 300 lines from the U.S. Forest Service

- 60 packets of tomato from D. W. Skrdla, Regional Plant Introduction Station, Ames, Iowa

- 20 packets of soybeans from R. Bernard, U.S. Regional Soybean Laboratory, Urbana, Illinois

- 10 packets of muskmelon from G. W. Bohn, ARS, Brawley, California

- 10 packets of vegetable seeds and clover assembled by A. Kelman, University of Wisconsin

- 10 packets of tobacco from J. R. Stavely, ARS, Belstville, Maryland

- 5 packets of tomato, onion, and cucumber seed from H. Munger, Cornell University

- potato seed (crosses) from J. Niederhauser, International Potato Center, Peru

- one packet of Proso millet from G. Burton, Tifton, Georgia

- 20 varieties of rice (4 sets) from N. Brady, IRRI, Philippines

- assorted lines of vegetables from the Asian Vegetable Research and Development Center, Taiwan

- 20 trees and other ornamentals from J. L. Creech, National Arboretum, Washington, D.C., in part presented by Longwood Gardens, Kennet Square, Pennsylvania, and William Flomer, Princeton Nursery, Princeton, New Jersey

- 50 herbarium specimens from T. G. Meyer, National Arboretum Herbarium, Washington, D.C.

Arrangements for shipment of the seed from the U.S. were made through the office of Mr. Harold Champeau, Agricultural Officer, American Consulate, Hong Kong. The materials were picked up by the delegation in Hong Kong and then hand-carried to Peking. It should be noted that the quantities of seeds and publications were of sufficient size that a large trunk had to be purchased in Hong Kong to accommodate these materials. The trunk came into good use throughout the China trip as it became the repository of seeds and publications presented by Chinese scientists.

Germ Plasm Received

The Chinese followed a pattern of presenting a collection of seeds on the departure of the U.S. delegation from each region visited. In some instances, items specifically requested were available but in most instances, seed consisted of lines selected by Chinese scientists. They are listed in order of locations visited.

Seeds presented by the Kirin Academy of Agricultural Sciences, September 5, 1974 (41 packets with descriptions):

- rice - 6 varieties	- sorghum - 6 lines
- soybeans - 11 varieties	- millet - 5 lines
- corn - 7 inbred lines	- spring wheat - 6 lines

Seeds presented by the Association of Agriculture and Forestry, Peking, September 6, 1974 (72 packets with descriptions):

- wheat - 8	- head cabbage - 2
- rice - 8	- stem lettuce - 1
- corn - 6	- eggplant - 3
- millet - 5	- string bean - 2
- oats - 6	- kidney bean - 1
- soybeans - 9	- turnip - 2
- tobacco - 3	- cucumber - 1

- tomato - 3   - luffa - 1

- Chinese cabbage - 2   - winter melon - 1

Seeds presented by the Northwest Agricultural College, Sian, Shensi Province, September 10, 1974 (7 packets with varietal names):

- wheat - 2

- corn - 4

- soybeans - 1

Seeds presented by the Kiangsu Academy of Agricultural Sciences Research Institute, Nanking, September 14, 1974 (28 packets with descriptions):

- rice - 4   - radish - 1

- wheat - 2   - onion - 2

- cotton - 1   - brassica narioosa - 1

- Chinese cabbage - 1   - Ai Chiao Huang - 1

- thorny cucumber - 1   - vetch (early) - 1

- tomato - 1   - wan-tou vetch - 1

Seeds presented by the Shensi Association of Agriculture, Sian, Shensi, September 11, 1974 (42 packets with varietal names):

- wheat - 8   - pepper - 1

- corn - 10   - gourd - 1

- Chinese cabbage - 2   - winter melon - 1

- cabbage - 1   - chestnut - 2

- common bean - 3   - jujube - 2

- celery - 1   - Crotalaria juncia - 1

- eggplant - 2   - alfalfa - 1

- tomato - 1   - soybean - 1

- cucumber - 2

Seeds presented by the Shanghai Association of Agriculture and Forestry, September 18, 1974 (18 packets with descriptions):

- rice - 3   - onion - 1

- 198 -

- barley - 2   - eggplant - 1

- rape - 1   - Dianthus - 1

- tomato - 1   - Althaea - 1

- pepper - 1   - Primula - 1

- Chinese cabbage - 2   - Cyclamen - 1

- garlic - 1   - Malva - 1

Seeds presented by the Kwangtung Association of Agriculture and
Forestry, September 22, 1974 (17 packets with descriptions):

- rice - 6   - Pinus bungeana - 1

- pea - 1   - Pinus tabulaeformis - 1

- melon - 2   - Thuja orientalis - 1

- cucumber - 2   - Sophora japonica - 1

- cowpea - 1   - Morus alba - 1

In addition, the following cuttings were presented by the South China
Botanical Garden, Canton:

- Carmona microphylla   - Haemaria discolor

- Buxus microphylla (wild)   - Begonia sp.

- Asarum sp.   - Ardisia mamellata

In addition, 50 herbarium specimens of native Chinese trees and shrubs
were presented by the Institute of Botany, Academy of Science, Peking, on September 2, 1974. An additional lot of specimens of trees and shrubs was presented by
the Northwest College of Agriculture and Forestry, Sian, Shensi. The living plants
were sent directly by airmail to the Plant Inspection Station, Washington, D.C., for
propagation at the National Arboretum.

All seeds received were accompanied by names, with some descriptions in
Chinese. These names and descriptions were translated by Mr. Alexander DeAngelis,
National Academy of Sciences, Secretary for the delegation. In many cases, variety
names in the detailed descriptions were translated when it was meaningful. Otherwise only transliterations were made. In all, a total of 231 packets of seeds or
lots of living plants were received by the delegation during the stay in China.
In addition, members of the delegation made roadside collections on occasion,
largely soybeans, turfgrasses, and ornamental shrubs.

- 199 -

In accordance with USDA quarantine regulations, all seeds and plants presented were sent from Hong Kong to the USDA Plant Inspection Station for treatment, inventory, and distribution, except for rice. All rice samples were sent to the International Rice Research Institute for increase and distribution. Wheat samples were divided at the Inspection Station and one-half of each sample was sent to CIMMYT, Mexico; the remainder went to Dr. J. Craddock, Agricultural Research Center, Beltsville. Soybeans will be increased by Dr. Bernard at Urbana, Illinois. The vegetables and forages will be handled through the regional intro-duction stations. The ornamental plants and seed were sent to the National Arb-oretum for propagation. Any quarantined materials such as corn, sorghum, and millet will have to be handled in accordance with USDA regulations. Inquiries concerning the descriptions and disposition of materials can be obtained by writing to Mr. Howard Hyland, Plant Introduction Office, Agricultural Research Center, Beltsville, Maryland 20705, or to members of the U.S. Delegation.

Because of short notice, the Chinese scientists promised to fill other requests such as the forest and tree seeds, soybeans, and vegetative materials at a later date.

Generally, except for inbreds, most of the materials received are peasant selections or materials from experiment stations which are known as "popular varieties," that is, those which have been put into use by commune farmers.

Requests from the U.S. Scientific Community and from Chinese Scientists

In preparation for the trip, American scientists were invited to submit requests to delegation members for Chinese plant materials. These were assembled and distributed to the delegation. It was apparent that our needs for Chinese germ plasm are far more complex than the Chinese scientists could cope with and requests were described only in rather general terms. We did indicate that many American scientists would be interested in Chinese contacts and we believe the list of some 200 Chinese scientists and administrators in China will serve as valid individuals to whom scientists may wish to write. A long list of requested forest and other tree species native to China was left in Peking for consideration. It remains to be seen to what extent this request will be filled.

The Chinese scientists in the several institutions visited submitted lists of materials they would like to receive. Delegation members agreed to assume responsibility for assembling materials that are available. These will be

sent to the requesting institution through the USDA Inspection Station, with
a copy of each shipping list forwarded to the Association of Agriculture and
Forestry, Peking.

Appendix 6

PUBLICATIONS RECEIVED

<u>Periodicals</u>

<u>Advances in Biochemistry and Biophysics</u>

    1974 (1) 5 copies

        (2) 5 copies

<u>Acta Botanica Sinica</u>

    1973 (1) and (2) 1 copy each

    1974 (1)   5 copies

        (2)   5 copies

<u>Acta Entomologica Sinica</u>

    1974 (1) 5 copies

        (2) 5 copies

        (3) 5 copies

<u>Acta Genetica Sinica</u>

    1974 (1) 5 copies

<u>Acta Phytotaxonimica Sinica</u>

    1973 (4) 1 copy

    1974 (1) 6 copies

        (2) 6 copies

        (3) 6 copies

<u>Acta Zoologica Sinica</u>

    1974 (1) 5 copies

        (2) 5 copies

<u>Plant Sciences Magazine</u>

    1974 #2 (Chinese Plant Society)

    1974 #1

Shih Sheng-han. 1974. A preliminary survey of the book Ch'i Min Yao Shu,
an agricultural encyclopedia of the 6th century. Science Press, Peking pp. 107.

Shih Sheng-han. 1974. On "Fan Sheng-Chih Shu" an agricultural book of China
written by Fan Sheng-chih in the first century B.C. Science Press, Peking pp. 68.

(Tomato cultivation in the northeast) Tung-pei fan-ch'ieh tsai-p'ei. Wang Hai-
t'ing author, pub. by K'e-hseuh ch'u-pan she (Scientific Press), 1972, 175 pp.
Chinese.

(Introduction to the important vegetable varieties of Peking City), Pei-ching-
shih chu-yao sh'u-ts'ai p'in-chung chieh-shao, Peking City Agricultural Sciences
Research Institute, pub. by Nung-ye ch'u-pan she (Agricultural Press) 1973, 239
pp. Chinese.

(Practices for high yields in corn) Yu-mi kao-ch'an ti shih - chien Shansi Province,
Hun-yuan County, Shih-yi-hao People's Commune, She-yi-hao Brigade Party Branch
Bureau, pub. by K'e-hsueh ch'u-pan she (Scientific Press), 1972, 39 pp.
Chinese.

(Revolutionary practices for high yields in cotton) Mien-hua kao-ch'an ti ke-
ming shih-chien, Honan Province, Hsin-hsiang County, Ch'i-li-ying Commune
Revolutionary Committee, pub. by K'e-hsueh ch'u-pan-she (Scientific Press), 1972,
148 pp. Chinese.

(Plant quarantine subjects) Chih-wu chien-yi tui-hsiang, Handbooks for Plant
Protection. Workers #7, Shanghai: Jenmin ch'u-pan she (People's Press), 20 pp.
and 15 pp. illustration, 1971. Chinese.

(Biological advances) Sheng-wu ti chin-hua, Fang Tzung-hsi author, K'e hsueh
ch'u-pan she (Scientific Press) 1973, 141 pp. Chinese.

(Advances and developments in the chemical initiation of biological nitrogen
fixation) Hua-hsueh mo-ni sheng-wu ku-tan chin-chan; K'e-hsueh ch'u-pan-she
(Scientific Press) compiled by the Group on Nitrogen Fixation of the Department
of Chemistry of Kirin University, 1973, 241 pp. Chinese.

(A collection of essays on plant introduction and acclimatization) Chih-wu
yin-chung hsun-hua lun-wen-chi, Chinese Academy of Sciences, Botanical Research
Institute, Peking Botanical Gardens, K'e-hsueh ch'u-pan she (Scientific Press)
1965. Chinese.

(Virus diseases of mushrooms and their prevention and cure) Shih chun ti chi
ch'i fang-chih, K'e-hsueh ch'u-pan she (Scientific Press) 1974, 62 pp. Chinese.

(Frequently seen and frequently used fungi) Ch'ang-chian yu ch'ang-yung chen-chun,
K'e-hsueh ch'u-pan she (Scientific Press), Chinese Academy of Sciences,
Institute of Microbiology, 1973, 317 pp. Latin index. Chinese.

Flora Illustrata Plantarum Primarium Sinicarum - Gramineae, Edit. Keng Yi-li,
1965, 1178 pp. Latin index, Nanking University, Biology Department and Chinese
Academy of Sciences Botanical Research Institute. Chinese.

(China's fungi) Chung-kuo ti chen-chun, Teng Shu-ch'un author, Science Press, 1964, 808 pp. Latin Index. Chinese.

(Publications from the conference on micro-nutrient deficiencies in plants) Chinese Academy of Sciences, Science Press, 1964, conference took place Dec. 10-15, 1962, 262 pp. Chinese.

(The control of rice pests) Handbook for Plant Protection Workers #1, 66 pp. Shanghai People's Press, 1974. Chinese.

(The common pesticides) Handbook for Plant Protection Workers #8, 86 pp. Shanghai People's Press, 1973. Chinese.

(The control of cotton pests) Handbook for Plant Protection Workers #2, 1973, 41 pp. Shanghai People's Press. Chinese.

(The breeding of micro-organisms through induced mutation), 83 pp. Science Press, 1973. Chinese.

(The control of pests of fruit trees) Handbook for Plant Protection Workers #6, 32 pp. 1972, Shanghai People's Press. Chinese.

(Plant quarantine subjects) Pamphlet #7, Shanghai People's Press. 20 pp.

(The control of pests of semi-arid crops and soybeans) Plant Protection Handbook #4, 26 pp. Shanghai People's Press, 1973. Chinese.

(Mass forecast of crop pests) 317 pp. Shanghai People's Press, 1973. Chinese.

(The control of vegetable pests) Plant Protection Handbook #5, 33 pp. Shanghai People's Press. 1973. Chinese.

(The pests of cotton and flax) Illustrated manual of pests of China's crops. Part 4, 93 pp. Agricultural Press, 1972, Chinese.

(The fungal diseases of cultivated plants in Kirin Province), P. K. Chih, J. H. Pai, G. H. Ch'i 366 pp. Science Press, 1966. Chinese.

Iconographia Cormophytorum Sinicorum, 2 vol., compiled by the Institute of Botany of the Chinese Academy of Sciences, 1972. Science Press, Vol. 1/1157 pp.; Vol. 2/1312 pp. Latin and Chinese index. Chinese.

(Rice pests and their control) edit. by Chou Ch'i, Shanghai People's Press, 1964 and 1972, 231 pp. Chinese.

(Struggle against potato degeneration) Chinese Academy of Sciences, Institute of Genetics, Tuber Group, revised edition, 1973. Science Press, 105 pp. Chinese.

(Plant better rice for revolution) Chinese Academy of Agriculture, Crop Breeding and Cultivation Research Institute, Agricultural Press, 1971. 144 pp. Chinese.

(Abundant yields in wheat: selection of model experiences) Agricultural Press, 1973, 102 pp. Chinese.

(The electro-chemical characteristics of soils and their research methods) Yu T'ien-jen etal. 1965, Science Press, 578 pp. Chinese.

(Control of diseases and insect pests in wheat, rape and green manure) Plant Protection Handbook #3, Shanghai People's Press, 1973, 58 pp. (includes 19 pp. of illustrations). Chinese.

Tachai: Standard Bearer in China's Agriculture, Peking, Foreign Language Press, 1972, 29 pp. English.

Serving the People with Dialectics: Essays on the Study of Philosophy by Workers and Peasants, Peking: Foreign Language Press, 1972, 48 pp. English.

On "Fan Sheng-chih Shu" an Agriculturalist Book of China written in the First Century B.C., Peking: Science Press, 1974, 68 pp., translated and commented upon by Shih Sheng-han, Ph.D., Professor of Plant Physiology, Northwest College of Agriculture. English.

A Preliminary Survey of the Book Ch'i Min Yao Shu, an Agricultural Encyclopaedia of the 6th Century, Peking, Science Press, 1974, 103 pp. Shih Sheng-han, Professor of Plant Physiology, Northwest College of Agriculture. English.

Flora Hainanica, Chinese Academy of Sciences, South China Botanical Garden, Vol. 1, Peking: Science Press, 1964, 517 pp. Chinese and Latin index. Chinese.

(Illustrated index of diseases and insect pests on China's agricultural crops: pamphlet #2 - wheat, barley, oats, rye, etc. diseases and insects) Agricultural Press, 86 pp. 1974. Chinese.

(A collection of vegetable production experiences in Peking) Peking: People's Press, 1974, 126 pp. Chinese.

SELECTED REFERENCES

Crook, Frederick W. and Linda A. Bernstein. 1974. Agriculture in the United
States and the People's Republic of China, 1967-1971. Foreign Agricultural
Economic Report No. 94, Economic Research Service, United States Department
of Agriculture (with 58 references).

Crook, Frederick W. and Linda A. Bernstein. 1974. The Agricultural situation
in the People's Republic of China and other Asian Communist countries, Review
of 1973 and Outlook for 1974. Economic Research Service, ERS - Foreign 362,
U.S. Department of Agriculture.

Ensminger, M.E. and Audrey Ensminger. 1973. China--The Impossible Dream.
Agriservices Foundation, Clovis, California.

Etienne, Gilbert (ed.). 1974. China's Agricultural Development. Asian Documentation and Research Center, Geneva, Switzerland.

Henle, H.V. 1974. Report on China's Agriculture. FAO, Rome.

Phillips, Christopher H. 1974. China's Agriculture, in U.S.-China Business
Review 1:2, pp. 36-52.

Phillips, Christopher H. 1974. Sino-U.S. Trade Statistics, 1972, 1973 including
agricultural trade. Special report No. 7, National Council for U.S.-China
Trade, Washington, D.C.

Phillips, Christopher H. 1974. The first annual report of the National Council
for U.S.-China Trade, (1100 17th Street N.W., Washington, D.C. 20036).

Phillips, Christopher H. The Technical Transformation of Agriculture in
Communist China. Praeger.

Scientific visit to the People's Republic of China. Research Branch, Agriculture Canada (mimeographed, 21 pages).

Signer, Ethan and Arthur W. Galston, 1972. Education and science in China.
Science, 175: No. 4017 pp. 15-23.

Stavis, Benedict, 1974. Making green revolution: The politics of agricultural
development in China. Rural Development Monograph No. 1, Rural Development
Committee, Cornell University (numerous references).

Swamy, Subramanian, 1973. Economic growth in China and India, 1952-70: A
comparative appraisal. Economic Development and Cultural Change 21:4 (II).

Tregear, T.R. 1965. The Geography of China, University of London Press.